SODIUM INTAKE IN POPULATIONS

ASSESSMENT OF EVIDENCE

Committee on the Consequences of Sodium Reduction in Populations

Food and Nutrition Board
Board on Population Health and Public Health Practice

Brian L. Strom, Ann L. Yaktine, and Maria Oria, *Editors*

INSTITUTE OF MEDICINE
OF THE NATIONAL ACADEMIES

THE NATIONAL ACADEMIES PRESS
Washington, D.C.
www.nap.edu

THE NATIONAL ACADEMIES PRESS   500 Fifth Street, NW   Washington, DC 20001

NOTICE: The project that is the subject of this report was approved by the Governing Board of the National Research Council, whose members are drawn from the councils of the National Academy of Sciences, the National Academy of Engineering, and the Institute of Medicine. The members of the committee responsible for the report were chosen for their special competences and with regard for appropriate balance.

This study was supported by Contract/Grant No. 200-2011-38807 between the National Academy of Sciences and the Centers for Disease Control and Prevention. Any opinions, findings, conclusions, or recommendations expressed in this publication are those of the author(s) and do not necessarily reflect the views of the organizations or agencies that provided support for the project.

International Standard Book Number-13: 978-0-309-28295-6
International Standard Book Number-10: 0-309-28295-0

Additional copies of this report are available for sale from the National Academies Press, 500 Fifth Street, NW, Keck 360, Washington, DC 20001; (800) 624-6242 or (202) 334-3313; http://www.nap.edu.

For more information about the Institute of Medicine, visit the IOM home page at: **www.iom.edu.**

The serpent has been a symbol of long life, healing, and knowledge among almost all cultures and religions since the beginning of recorded history. The serpent adopted as a logotype by the Institute of Medicine is a relief carving from ancient Greece, now held by the Staatliche Museen in Berlin.

Suggested citation: IOM (Institute of Medicine). 2013. *Sodium intake in populations: Assessment of evidence.* Washington, DC: The National Academies Press.

"Knowing is not enough; we must apply.
Willing is not enough; we must do."
—Goethe

INSTITUTE OF MEDICINE
OF THE NATIONAL ACADEMIES

Advising the Nation. Improving Health.

THE NATIONAL ACADEMIES
Advisers to the Nation on Science, Engineering, and Medicine

The **National Academy of Sciences** is a private, nonprofit, self-perpetuating society of distinguished scholars engaged in scientific and engineering research, dedicated to the furtherance of science and technology and to their use for the general welfare. Upon the authority of the charter granted to it by the Congress in 1863, the Academy has a mandate that requires it to advise the federal government on scientific and technical matters. Dr. Ralph J. Cicerone is president of the National Academy of Sciences.

The **National Academy of Engineering** was established in 1964, under the charter of the National Academy of Sciences, as a parallel organization of outstanding engineers. It is autonomous in its administration and in the selection of its members, sharing with the National Academy of Sciences the responsibility for advising the federal government. The National Academy of Engineering also sponsors engineering programs aimed at meeting national needs, encourages education and research, and recognizes the superior achievements of engineers. Dr. C. D. Mote, Jr., is president of the National Academy of Engineering.

The **Institute of Medicine** was established in 1970 by the National Academy of Sciences to secure the services of eminent members of appropriate professions in the examination of policy matters pertaining to the health of the public. The Institute acts under the responsibility given to the National Academy of Sciences by its congressional charter to be an adviser to the federal government and, upon its own initiative, to identify issues of medical care, research, and education. Dr. Harvey V. Fineberg is president of the Institute of Medicine.

The **National Research Council** was organized by the National Academy of Sciences in 1916 to associate the broad community of science and technology with the Academy's purposes of furthering knowledge and advising the federal government. Functioning in accordance with general policies determined by the Academy, the Council has become the principal operating agency of both the National Academy of Sciences and the National Academy of Engineering in providing services to the government, the public, and the scientific and engineering communities. The Council is administered jointly by both Academies and the Institute of Medicine. Dr. Ralph J. Cicerone and Dr. C. D. Mote, Jr., are chair and vice chair, respectively, of the National Research Council.

www.national-academies.org

COMMITTEE ON THE CONSEQUENCES OF
SODIUM REDUCTION IN POPULATIONS

BRIAN L. STROM (*Chair*), George S. Pepper Professor of Public Health, University of Pennsylvania School of Medicine, Philadelphia

CHERYL A. M. ANDERSON, Associate Professor of Family and Preventive Medicine, University of California, San Diego

JAMY ARD, Associate Professor of Epidemiology and Prevention, Wake Forest Baptist Health, Winston-Salem, North Carolina

KIRSTEN BIBBINS-DOMINGO, Associate Professor of Medicine and of Epidemiology and Biostatistics, University of California, San Francisco, and Co-Director, San Francisco General Hospital, California

NANCY R. COOK, Professor in the Department of Medicine at Harvard Medical School and Brigham & Women's Hospital, Boston, Massachusetts

MARY KAY FOX, Senior Researcher, Mathematica Policy Research, Inc., Cambridge, Massachusetts

NIELS GRAUDAL, Senior Consultant, Copenhagen University Hospital, Rigshospitalet, Denmark

JIANG HE, Joseph S. Copes Professor of Epidemiology, Tulane University School of Public Health and Tropical Medicine, New Orleans, Louisiana

JOACHIM IX, Associate Professor of Medicine, Veterans Affairs San Diego Healthcare System, California

STEPHEN E. KIMMEL, Professor of Medicine and Epidemiology, University of Pennsylvania School of Medicine, Philadelphia

ALICE H. LICHTENSTEIN, Gershoff Professor of Nutrition Science and Policy, Tufts University, Boston, Massachusetts

MYRON WEINBERGER, Professor Emeritus of Medicine, Indiana University School of Medicine, Indianapolis and Editor-in-Chief, *Journal of the American Society of Hypertension*

IOM Staff

MARIA ORIA, Study Director
ANN L. YAKTINE, Study Director
JULIA HOGLUND, Research Associate
COLIN F. FINK, Senior Program Assistant
ANTON BANDY, Financial Officer
GERALDINE KENNEDO, Administrative Assistant
LINDA D. MEYERS, Director, Food and Nutrition Board
ROSE MARIE MARTINEZ, Director, Board on Population Health and Public Health Practices

Reviewers

This report has been reviewed in draft form by individuals chosen for their diverse perspectives and technical expertise, in accordance with procedures approved by the National Research Council's Report Review Committee. The purpose of this independent review is to provide candid and critical comments that will assist the institution in making its published report as sound as possible and to ensure that the report meets institutional standards for objectivity, evidence, and responsiveness to the study charge. The review comments and draft manuscript remain confidential to protect the integrity of the deliberative process. We wish to thank the following individuals for their review of this report:

MICHAEL H. ALDERMAN, Albert Einstein College of Medicine
DAVID B. ALLISON, University of Alabama, Birmingham
LAWRENCE J. APPEL, Johns Hopkins University
GLENN M. CHERTOW, Stanford University School of Medicine
JOHANNA T. DWYER, Tufts Medical Center
SHIRIKI K. KUMANYIKA, University of Pennsylvania Perelman School of Medicine
JOSEPH LAU, Brown University
DAVID A. McCARRON, University of California, Davis
SUZANNE P. MURPHY, University of Hawaii, Manoa
SUZANNE OPARIL, University of Alabama, Birmingham
DONALD B. RUBIN, Harvard University
ANNA MARIA SIEGA-RIZ, University of North Carolina at Chapel Hill
JUDITH S. STERN, University of California, Davis

Although the reviewers listed above have provided many constructive comments and suggestions, they were not asked to endorse the conclusions or recommendations nor did they see the final draft of the report before its release. The review of this report was overseen by **LYNN R. GOLDMAN,** George Washington University, and **SUSAN J. CURRY,** University of Iowa. Appointed by the National Research Council and the Institute of Medicine, they were responsible for making certain that an independent examination of this report was carried out in accordance with institutional procedures and that all review comments were carefully considered. Responsibility for the final content of this report rests entirely with the authoring committee and the institution.

Preface

Heart disease and stroke are major cardiovascular diseases (CVDs) and leading causes of death for both men and women in the United States. Major risk factors of CVD, such as high blood pressure, high levels of low-density lipoprotein cholesterol, and smoking, are frequently found among the U.S. population. Other factors, however, such as a poor diet, also can contribute to high blood pressure and disease risk. For more than four decades, controversies have surrounded the contribution of sodium consumption as a risk factor for noncommunicable diseases, including CVD. Numerous domestic and international organizations and governments have advised populations against consuming high levels of sodium. Investigators have applied models based on blood pressure decreases to predict the number of deaths that would be saved in the general population from reducing sodium consumption. Although most biomarkers have limitations as indicators of risk of adverse health outcomes, the evidence for blood pressure as a surrogate endpoint for risk of CVD and stroke is widely recognized and accepted. Sodium, however, might execute its functions through various processes in addition to blood pressure. Thus, any strategy to reduce the consumption of sodium should include studies that demonstrate an association between sodium consumption and direct health effects. Although there is agreement that sodium policies should be guided by research based on direct health outcomes, unfortunately, randomized controlled trials and observational studies that have looked at the association between sodium and direct health outcomes are few and have been interpreted in many different ways. These differences underscore the difficulties in accurately measuring the long-term health effects of diets or individual nutrients that are consumed in the context of a diet.

In 2005, as part of its work on Dietary Reference Intakes, the Institute of Medicine (IOM) conducted a review of the scientific literature on the association between sodium and health effects, including both intermediate and direct health outcomes. Since that time, more data have been collected but the controversies continue and have slowed down the ability to implement sodium policies that are consistent with the current *Dietary Guidelines for Americans*. Adding to these longstanding controversies is emerging evidence on potential adverse effects of a too-low dietary sodium intake for some population subgroups. The need for an in-depth examination of the scientific literature became obvious. The Centers for Disease Control and Prevention of the U.S. Department of Health and Human Services asked the IOM to convene an expert committee to examine the designs, methodologies, and conclusions of this emerging evidence and to comment on the implications of their findings for population-based sodium-reduction strategies. Of particular interest to the agency was assessing benefits and adverse outcomes of reducing sodium intake in the population and in relevant subgroups that have been described as being particularly at risk. These subgroups are individuals with hypertension and prehypertension, those 51 years of age and older, African Americans, and those with diabetes, chronic kidney disease, and congestive heart failure. The committee conducted an extensive review of the peer-reviewed literature. A public workshop also was conducted, which provided an opportunity for experts outside the committee to present some of the most controversial data and for stakeholders to describe their positions on relevant issues. The contributions of the workshop speakers and additional information gathered by the committee were invaluable for its deliberations.

I am deeply indebted to the committee members who generously gave their time and effort to complete this task in a very short time. Their diversity in expertise and backgrounds and in-depth discussions shed light on highly complex scientific issues. In addition, on behalf of the committee, I would like to thank the IOM staff, study co-directors Ann L. Yaktine and Maria Oria, research associate Julia Hoglund, and senior program assistant Colin Fink, who worked tirelessly to complete this task. My gratitude also goes to the Director of the Food and Nutrition Board, Linda D. Meyers, and the Director of the Board on Population Health and Public Health Practices, Rose Marie Martinez, for their guidance and support during the entire study.

It is my hope that the conclusions of this committee will add to current discussions about sodium intake and health and will help future policy makers as they continue to decide and implement sodium strategies that will benefit public health.

Brian L. Strom, *Chair*
Committee on the Consequences of
Sodium Reduction in Populations

Contents

Summary[1]

Despite efforts over the past several decades to reduce sodium intake in the United States, adults still consume an average of 3,400 milligrams (mg) of sodium every day. A number of scientific bodies and professional health organizations, including the American Heart Association, the American Medical Association, and the American Public Health Association, support reducing dietary sodium intake, and the 2010 *Dietary Guidelines for Americans* includes as a goal to "reduce daily sodium intake to less than 2,300 milligrams (mg) and further reduce intake to 1,500 mg among persons who are 51 years of age and older and those of any age who are African American or have hypertension, diabetes, or chronic kidney disease."

A substantial body of evidence supports these efforts to reduce sodium intake. This evidence links excessive dietary sodium to high blood pressure, a surrogate marker for cardiovascular disease (CVD), stroke, and cardiac-related mortality. However, concerns have been raised that a low sodium intake may adversely affect certain risk factors, including blood lipids and insulin resistance, and thus potentially increase risk of heart disease and stroke. In fact, several recent reports have challenged sodium reduction in the population as a strategy to reduce this risk.

---

[1]This summary does not contain references. Citations to support statements made herein are given in the body of the report.

COMMITTEE TASK AND APPROACH

Against the backdrop of questions about sodium reduction in the population, the Centers for Disease Control and Prevention of the U.S. Department of Health and Human Services (HHS) asked the Institute of Medicine (IOM) to convene an expert committee to examine the designs, methodologies, and conclusions of this emerging evidence, as well as other reports published since the 2005 Dietary Reference Intake (DRI) report, *Dietary Reference Intakes for Water, Sodium, Chloride, and Sulfate*. Specifically, the committee was asked to review and assess the benefits and adverse outcomes (if any) of reducing sodium intake in the population, particularly in the range of 1,500 to 2,300 mg per day, with an emphasis on relevant subgroups. These subgroups include individuals with hypertension and prehypertension, those 51 years of age and older, African Americans, and those with diabetes, chronic kidney disease, and congestive heart failure (CHF). The committee also was asked to comment on the implications for population-based strategies to reduce sodium intake and to identify data and methods gaps and suggest ways to address them.

In approaching its task, the committee first conducted a broad search of the published literature through 2012 to identify relevant scientific publications on sodium and direct health outcomes. The committee was unable to identify studies published prior to 2003 that provided data on how the frequency of direct health outcomes was associated with changes in dietary sodium. Information also was gathered from a public workshop. The committee then developed a strategy to qualitatively assess each study identified as relevant from the search. Although an in-depth review was not conducted, the committee also considered evidence published since 2003 on associations between sodium intake and intermediate markers, particularly blood pressure. This additional evidence on the effect of sodium on blood pressure supported the findings and conclusions in the DRI report, the technical report from the 2010 Dietary Guidelines Advisory Committee (DGAC), and the recently released 2012 report from the World Health Organization, *Guideline: Sodium Intake for Adults and Children*.

Although blood pressure is a widely accepted surrogate marker, the scientific community continues to debate the use of other intermediate markers and biomarkers generally. Further, and in keeping with the systematic evidence review in the DGAC report, the effects of lowering sodium intake on blood pressure, as with other biomarkers, cannot always be disentangled from the effects of total dietary modification. For example, the committee's review revealed that in a number of studies, the effects of dietary sodium on CVD outcomes sometimes persisted even after controlling for blood pressure. This suggests that associations between dietary sodium and risk of CVD may be mediated through interaction with other dietary factors (e.g., the effects of other electrolytes), or through pathways in addition to blood pressure. The committee sought to synthesize these potential associations. The committee's approach to assessing the evidence focused on new data about the health effects of sodium intake on measures of health outcomes, rather than on effects mediated through an intermediate marker, namely blood pressure (see Figure D-1 in Appendix D).

The committee's assessment of the evidence reviewed was guided by a number of factors. These included the study design, the quantitative measures of dietary sodium intake and confounder adjustment, as well as the number and consistency of relevant studies available. Assessing the impact of sodium intake on health outcomes was complicated by variability in the types and quality of measures used in observational studies, so that measures could not be reliably calibrated across studies. These measures also were difficult to assess in comparison to sodium intake in clinical trials. It was the consensus of the committee that the lack of consistency among studies in the methods used for defining sodium intakes at both high and low ends of the range of typical intakes among various population groups precluded deriving a numerical definition for high and low intakes in its findings and conclusions. Likewise, the extreme variability in intake levels between and among population groups precluded the committee from establishing a "healthy" intake range. The committee could consider sodium intake levels only within the context of each individual study.

The evidence for an effect of sodium intake on health outcomes reviewed by the committee included a broad range of population groups and methodological approaches. All of the evidence on the health outcomes related to CVD, stroke, and mortality was observational, mostly prospective cohort studies, whereas the evidence on health outcomes related to heart failure included randomized clinical trials (RCTs). Although the committee considered using a meta-analysis to assess the evidence, this approach was deemed inappropriate for this review because of the marked heterogeneity among the reviewed studies, particularly with respect to variations in measuring sodium intake and adjusting for confounders. For the same reason, the committee did not use a rating system to evaluate individual studies. Instead, studies were reviewed and assessed on an individual basis, and the committee considered the evidence on associations between sodium intake and health outcomes in its totality.

In evaluating each study, the committee considered RCTs a higher-quality study design for determining the effect of sodium on health outcomes than were observational studies. Well-executed cohort studies were considered more important for suggesting associations between dietary intake and health outcomes than were case-control studies because of the potential for bias in dietary assessment of sodium intake with the case-control study design. Finally, cross-sectional studies were included as an indication of a potential association or to support (or not support) results from other studies. The committee used two major criteria to assess the quality of the evidence for all study designs: (1) the method to estimate sodium intake and the quality of its implementation, and (2) confounder adjustment. Other criteria included the approach used to change sodium intake, the instrument used to estimate sodium intake, the length of the intervention or follow-up of participants, interactions with other factors, and generalizability to the general population or population subgroups.

FINDINGS AND CONCLUSIONS

Recognizing the limitations of the available evidence, the committee found no consistent evidence to support an association between sodium intake and either a beneficial or adverse effect on most direct health outcomes other than some CVD outcomes (including stroke and CVD mortality) and all-cause mortality. Some evidence suggested that decreasing sodium intake could possibly reduce the risk of gastric cancer. However, the evidence was too limited to conclude the converse—that higher sodium intake could possibly increase the risk of gastric cancer. Interpreting these findings was particularly challenging because most studies were conducted outside the United States in populations consuming much higher levels of sodium than those consumed in this country. Thus, the committee focused its findings and conclusions on evidence for associations between sodium intake and risk of CVD-related events and mortality.

Findings and Conclusions for Cardiovascular Disease, Stroke, and Mortality

General U.S. Population

Finding 1: The committee found that the results from studies linking dietary sodium intake with direct health outcomes were highly variable in methodological quality, particularly in assessing sodium intake. The range of limitations included over- or underreporting of intakes or incomplete collection of urine samples. In addition, variability in data collection methodologies limited the committee's ability to compare results across studies.

Conclusion 1: Although the reviewed evidence on associations between sodium intake and direct health outcomes has methodological flaws and limitations, the committee concluded that, when considered collectively, it indicates a positive relationship between higher levels of sodium intake and risk of CVD. This evidence is consistent with existing evidence on blood pressure as a surrogate indicator of CVD risk.

Finding 2: The committee found that the evidence from studies on direct health outcomes was insufficient and inconsistent regarding an association between sodium intake below 2,300 mg per day and benefit or risk of CVD outcomes (including stroke and CVD mortality) or all-cause mortality in the general U.S. population.

Conclusion 2: The committee determined that evidence from studies on direct health outcomes is inconsistent and insufficient to conclude that lowering sodium intakes below 2,300 mg per day either increases or decreases risk of CVD outcomes (including stroke and CVD mortality) or all-cause mortality in the general U.S. population.

Population Subgroups

Finding 1: The committee found that the evidence from multiple RCTs that were conducted by a single investigative team indicated that low sodium intake (e.g., down to 1,840 mg per day) may lead to greater risk of adverse events in CHF patients with reduced ejection fraction and who are receiving certain aggressive therapeutic regimens. This association also is supported by one observational study in which low sodium intake levels in patients with CVD and diabetes were associated with higher risk of CHF events.

Conclusion 1: The committee concluded that the available evidence suggests that low sodium intakes may lead to higher risk of adverse events in mid- to late-stage CHF patients with reduced ejection fraction and who are receiving aggressive therapeutic regimens. Because these therapeutic regimens were very different than current standards of care in the United States, the results may not be generalizable. Similar studies in other settings and using regimens more closely resembling those in standard U.S. clinical practice are needed.

Finding 2: The committee found that data among prehypertensive participants from two related studies provided some evidence suggesting a continued benefit of lowering sodium intake in these patients down to 2,300 mg per day (and lower, although based on small numbers in the lower range). In contrast, the committee found no evidence for benefit and some evidence suggesting risk of adverse health outcomes associated with sodium intake levels in ranges approximating 1,500 to 2,300 mg per day in other disease-specific population subgroups, specifically those with diabetes, chronic kidney disease (CKD), or preexisting CVD. In addition to inconsistencies in sodium intake measures, methodological flaws included the possibility of confounding and reverse causality. No relevant evidence was found on health outcomes for other population subgroups considered (i.e., persons

51 years of age and older and African Americans). In studies that explored interactions, race, age, or prevalence of hypertension or diabetes did not change the effect of sodium on health outcomes.

Conclusion 2: The committee concluded that, with the exception of the CHF patients described above, the current body of evidence addressing the association between low sodium intake and health outcomes in the population subgroups considered is limited. The evidence available is inconsistent and limited in its approaches to measuring sodium intake. The evidence also is limited by small numbers of health outcomes and the methodological constraints of observational study designs, including the potential for reverse causality and confounding.

The committee further concluded that, while the current literature provides some evidence for adverse health effects of low sodium intake among individuals with diabetes, CKD, or preexisting CVD, the evidence on both the benefit and harm is not strong enough to indicate that these subgroups should be treated differently from the general U.S. population. Thus, the committee concluded that the evidence on direct health outcomes does not support recommendations to lower sodium intake within these subgroups to, or even below, 1,500 mg per day.

Implications for Population-Based Strategies to Gradually Reduce Sodium Intake in the U.S. Population

As noted in Chapter 1, recommendations of the Panel on Dietary Reference Intakes for Electrolytes and Water of an Adequate Intake for sodium of 1,500 mg per day for all individuals 9 years of age up to 51 years of age was set as an amount necessary to achieve an overall diet that provides an adequate intake of other nutrients and also covers sodium sweat losses. A Tolerable Upper Intake Level for sodium was set at 2,300 mg per day based on evidence showing associations between high sodium intakes and risk of high blood pressure and consequent risk of CVD, stroke, and mortality.

Given this background, overall, the committee found that the available evidence on associations between sodium intake and direct health outcomes is consistent with population-based efforts to lower excessive dietary sodium intakes, but it is not consistent with recommendations that encourage lowering of dietary sodium in the general population to 1,500 mg per day. Further, as noted in the 2010 DGAC report, population subgroups, including those with diabetes, CKD, or preexisting CVD, individuals with hypertension, prehypertension, persons 51 years of age and older, and African Americans represent, in aggregate, a majority of the general U.S. population. Thus, when considered in light of the current state of the evidence on associations between sodium intake and direct health outcomes for these subgroups, except when data specifically indicate they are different, there is not sufficient evidence to support treating them differently from the general U.S. population.

The committee was not asked to draw conclusions about a specific target range of dietary sodium for the general population or for population subgroups. However, the committee notes that there are important factors it considered that preclude such a conclusion. For example, one factor that is often discussed in the context of other health-related questions is the challenge of defining specific intake levels when the variables of interest are continuous. That is an especially difficult issue in the present circumstances, where the target intake level could theoretically differ for different, large population subgroups.

Other methodological factors that preclude making conclusions about a specific target range for sodium relate to the variability in approaches and study designs in the literature reviewed. Most importantly, quantitative methods for measuring dietary sodium intake have limitations and there are impediments to calibrating those measures across different methodological approaches and study designs. Methodological problems in assessing sodium intake make this particularly challenging.

FUTURE RESEARCH TO ADDRESS GAPS IN DATA AND METHODOLOGY

The committee identified a number of data and methods gaps in studies on sodium intake and risk of adverse health outcomes among population groups. Further research in the areas highlighted below would strengthen the evidence base on the association between lower (1,500 to 2,300 mg) levels of sodium and health outcomes in the general population and population subgroups:

1. Standardized methodological approaches to measure sodium intake in population groups. Specific examples include standardizing the use of 24-hour urine collections and validating sodium intake estimates with data on urine volume, urine creatinine, and body weight;

2. Approaches using dietary sodium intake levels corresponding to levels in current guidelines (i.e., 1,500 to 2,300 mg per day) when examining associations between sodium intake and health outcomes;

3. Analyses examining the effects on health outcomes of dietary sodium in combination with other electrolytes, particularly potassium;

4. Methods that account for confounding factors in dietary studies, including the influence of reported total daily caloric intake on observational associations between sodium and health outcomes, and methods that clarify attributes of individuals with apparently low sodium intake or excretion; and

5. Analyses of interactions with antihypertensive medication and blood pressure in studies examining associations between sodium intake and health outcomes.

In addition, the committee identified a need for RCT research, and observational and mechanistic studies, particularly in population subgroups. Examples of such clinical trials include those to examine

1. effects of a range of sodium levels on risk of cardiovascular events, stroke, and mortality among
   a. patients in controlled environments, where randomized trials may be more feasible, such as the elderly in chronic care facilities and other institutionalized individuals; and
   b. individuals as part of natural experiments, such as those in other countries where policies affecting sodium consumption are in effect;

2. effects of low-sodium diets on adverse events among CHF patients receiving therapeutic treatment modalities typically used in the United States; and

3. potential beneficial or adverse outcomes of a range of sodium intakes among African Americans, adults 51-70 years of age, 70 years of age and older, and other population subgroups; RCTs may be particularly important within higher-risk patient populations, where reverse causation is a potential limitation of observational studies.

The committee also identified a need for studies to collect and reanalyze

1. data from existing clinical trials that were designed to evaluate sodium and health; and

2. data during extended follow-up periods after completion of a clinical trial to identify health outcomes, such as mortality, that could manifest later in life and after longer follow-up periods. Such trials would not be simple to conduct, however, and careful feasibility assessment is needed first.

In addition to RCT research, mechanisitic studies are needed to examine potential physiologic changes associated with lowering sodium intake and adverse health outcomes. Finally, additional observational research is needed to examine associations between sodium intake and cancer, especially gastric cancer, in the U.S. population, as well as associations between sodium intake and caloric intake in both short-term and longitudinal studies.

1

Introduction

BACKGROUND

Sodium is an essential nutrient and, as the primary cation in extracellular fluid, is required for a number of physiologic activities, including maintenance of extracellular volume and plasma osmolality, cellular membrane potential, and functioning of active transport systems. About 95 percent of total sodium is in extracellular fluid. Sodium balance is regulated collectively by the renin-angiotensin-aldosterone system, the sympathic nervous system, atrial natriuretic peptide, the kalikrein-kinin system, intra-renal mechanisms, and other factors. The majority of ingested sodium (more than 90 percent) is excreted in the urine (unless sweating is excessive).

Previously established evidence strongly supports that consumption of excessive sodium is a risk factor for high blood pressure, which in turn is a strong risk factor for consequent health outcomes, including cardiovascular disease (CVD), stroke, and mortality. However, emerging evidence suggests sodium intakes below 2,300 milligrams (mg) per day may increase risk of adverse health outcomes, at least in some population subgroups. Thus, debate has emerged about the level of dietary sodium intake that is associated with risk of adverse outcomes. The following discussion briefly summarizes current recommendations for sodium intake for the healthy U.S. population and for at-risk populations, namely African Americans, those 51 years of age and older, and those with hypertension or prehypertension, diabetes, or chronic kidney disease (CKD).

The Institute of Medicine (IOM) Panel on Dietary Reference Intakes (DRIs) for Electrolytes and Water was charged with establishing DRIs for sodium. DRIs comprise a set of nutrient reference values for assessing and planning diets for healthy people. These reference values replace and expand upon the previous Recommended Dietary Allowances (RDAs) for the United States and the Recommended Nutrient Intakes for Canada. DRIs encompass RDAs (derived from the Estimated Average Requirements, EARs) and Tolerable Upper Intake Levels (ULs) for life stages and genders. Detailed information about the process of establishing DRIs can be found in the report *Dietary Reference Intakes: The Essential Guide to Nutrient Requirements* (IOM, 2006). A brief summary of the DRIs is found in Box 1-1.

In establishing DRI values for sodium, the Panel on Dietary Reference Intakes for Electrolytes and Water (IOM, 2005) found insufficient evidence to derive RDAs. Instead Adequate Intakes (AIs) were set for all life stage and gender groups. For example, an AI of 1,500 mg per day was set for all children 9 years of age and older, adolescents, and adult men and women up to 51 years of age to ensure that the overall diet provides an adequate intake of other important nutrients and also to cover sweat losses in unacclimated individuals who are exposed to high temperatures or who become physically active (IOM, 2006, p. 388). The AIs for those 51 to 70

BOX 1-1
Definition of Dietary Reference Intakes

Adequate Intake (AI): The recommended average daily intake level based on observed or experimentally determined approximations or estimates of nutrient intake by a group (or groups) of apparently healthy people that are assumed to be adequate. The AI is used when an RDA cannot be determined.

Estimated Average Requirement (EAR): The average daily nutrient intake level that is estimated to meet the requirements of half of the healthy individuals in a particular life stage and gender group.

Recommended Dietary Allowance (RDA): The average daily dietary nutrient intake level that is sufficient to meet the nutrient requirements of nearly all (97-98 percent) healthy individuals in a particular life stage and gender group.

Tolerable Upper Intake Level (UL): The highest average daily nutrient level that is likely to pose no risk of adverse health effects to almost all individuals in the general population. As intake increases above the UL, the potential risk of adverse effects may increase.

SOURCE: IOM, 2006.

years of age and 70 years or older were set at 1,300 and 1,200 mg per day, respectively. A UL was established for sodium based on evidence showing associations between excessive sodium intake and risk of high blood pressure and consequent risk of CVD, stroke, and mortality (see Chapter 3 for a detailed discussion). Evidence from intervention studies showed no threshold in the dose-response curve of the relationship between sodium intake level and blood pressure, and therefore a no-observed adverse effect level (NOAEL) could not be established. Instead, a UL of 2,300 mg per day for males and females 14 years and above was established based on the lowest observed adverse effect level (LOAEL) from the dose–response evidence for blood pressure. For some population subgroups, namely older individuals, African Americans, and those with hypertension, diabetes, or CKD, the report suggested the UL should be lower than 2,300 mg/day although a defined level was not determined (see IOM, 2005, pp. 387-394, for more detailed information).

The *Dietary Guidelines for Americans* (DGA) are the basis for federal nutrition policy and nutrition programs. The goals of the DGA have evolved since the first DGA were published in 1980, reflecting changes in the science of nutrition and understanding of the role of nutrition in health. Every 5 years the DGA are revised, drawing from a technical review of evidence by a panel of experts, the Dietary Guidelines Advisory Committee (DGAC). In 2005, the DGAC concluded that the general adult population reduce sodium intake to less than 2,300 mg of sodium per day (HHS and USDA, 2005). The committee further concluded that individuals with hypertension, African Americans, and middle-aged and older adults (51 years of age and older) would benefit from reducing their sodium intake even further (HHS and USDA, 2005).

Because these latter groups together now comprise a majority of U.S. adults, the 2010 DGAC technical report concluded that a goal for sodium intake reduction should be 1,500 mg per day for the general population (HHS and USDA, 2010a). The *Dietary Guidelines for Americans, 2010* (HHS and USDA, 2010b), the federal nutrition policy document, recommended retaining the goal that the general adult population should reduce consumption to less than 2,300 mg per day, but only that individuals 51 years of age and older, African Americans, and those with hypertension, diabetes, and CKD should reduce intake to 1,500 mg per day.

As with the DRI report (IOM, 2005), the rationale for the DGAC (HHS and USDA, 2010b) conclusion is grounded in evidence for a relationship between sodium intake and blood pressure as a surrogate marker of disease, and supported by additional evidence that decreasing blood pressure is associated with decreased risk of adverse health outcomes, particularly stroke and ischemic heart disease. Although the scientific community continues to debate the use of biomarkers in general and surrogate indicators of health outcomes, recent evidence attributes 35 percent of myocardial infarction and stroke events, 49 percent of heart failure episodes, and 24 percent of premature deaths to high blood pressure (Lawes et al., 2008).

Based on analyses of intake data from the 2003-2006 *What We Eat in America* dietary interview component of the National Health and Nutrition Examination Survey (NHANES), the 2010 DGAC report (HHS and USDA, 2010b) found that usual intakes of sodium were excessive for most age and gender groups in the U.S. population. Specifically, usual sodium intake exceeded the AI for more than 97 percent of all age and gender groups and exceeded the UL for more than 90 percent of boys older than 9 years of age and adult men up to 70 years of age. Among women, usual sodium intakes exceeded the UL in 84 and 75 percent of girls older than 9 years of age and women 19 years and older, respectively (ARS, 2010).

As a result of the persistent evidence that excessive sodium intake increases blood pressure, a risk factor for CVD, stroke, and mortality, federal nutrition policy now includes an emphasis on decreasing sodium intake in the general population as a preventive measure against risk of these adverse health outcomes. Based on the same evidence, various other domestic (e.g., American Heart Association) and international (e.g., World Health Organization) organizations have published recommendations for sodium consumption. Although they may differ from the DGA recommendations, they generally agree that sodium consumption is excessive among populations worldwide, and that it should be gradually reduced. As with the findings from the evidence reviews in IOM (2005) and DGAC (HHS and USDA, 2010a), these recommendations are grounded in evidence showing associations between excessive sodium intake and risk of high blood pressure and the strong association between hypertension and risk of CVD, stroke, and mortality.

Efforts to reduce sodium intake in the population have included educating the public about risks associated with excessive sodium consumption, and encouraging the food industry to reduce the sodium content of processed foods or develop alternate low-sodium products. The IOM report *Strategies to Reduce Sodium Intake in the United States* (IOM, 2010) reviewed indicators of dietary sodium intake and assessed a range of efforts aimed at reducing sodium consumption (e.g., consumer education, product labeling, and product reformulation) in both the public and private sectors. The report noted that consumers live in a food environment in which social, environmental, and macrolevel factors influence the types of foods consumed, highlighted the difficulties consumers face in consuming no more than 2,300 mg of sodium per day in this environment, and concluded that current strategies to reduce sodium consumption are ineffective. It also emphasized that the food supply is a key obstacle to reducing sodium consumption even when a myriad of strategies have been implemented.

To illustrate, NHANES 2009-2010 data show mean daily sodium intakes among U.S. adults 20 years of age and older are 2,980 and 4,243 mg for women and men, respectively. A salt adjustment is not applied to the 2009-2010 or later surveys; estimates of sodium intake include salt added in cooking and food preparation as assumed in the nutrient profiles for foods. These results are similar to the data from NHANES 2007-2008 (3,000 and 4,224 mg for women and men, respectively), indicating that sodium consumption patterns have changed little. A recent analysis of data from NHANES 2003-2008 showed that 90.7 percent (confidence interval [CI]: 89.6-91.8) of adults 20 years of age and older consumed more than 2,300 mg per day. Moreover, the same analysis showed that among U.S. subpopulations at risk (i.e., African Americans, those 51 years of age and older, or persons with hypertension, diabetes, or CKD), 98.8 percent (CI: 98.4-99.2) overall consumed more than 1,500 mg per day (Cogswell et al., 2012). Collectively, these findings pointed to the need for a gradual reduction in sodium consumption in the U.S. population.

THE COMMITTEE'S TASK

Despite the evidence that decreasing sodium intake reduces risk of high blood pressure, which in turn is strongly linked to CVD and other adverse health outcomes, new evidence has emerged that raises questions about potential adverse effects, specifically effects on insulin resistance and risk of CVD, associated with reducing sodium intake levels below 2,300 mg per day. Whether this evidence refutes previous evidence must be considered in light of the approaches used in these studies to examine health effects mediated directly through interactions with sodium, rather than examining blood pressure as a surrogate indicator of disease risk.

The controversy about possible adverse consequences to some individuals of low sodium intakes, and therefore of population strategies to reduce sodium consumption, prompted the Centers for Disease Control and Prevention (CDC) to ask the IOM to convene a committee to review the scientific evidence published since 2003 about potential benefits and adverse effects on health outcomes of sodium intake, particularly in the ranges of 1,500 to 2,300 mg per day. The committee also was asked to comment on the implications that this new evidence will have on population-based strategies to reduce sodium intake in the population and to identify research gaps. The statement of task is shown in Box 1-2. In its task, the committee was asked to consider potential benefits and adverse outcomes in the U.S. population generally and for relevant subsets of the population, namely those 51 years of age and older, African Americans, and those with hypertension, prehypertension, diabetes, CKD, and congestive heart failure.

The 2010 DGA (HHS and USDA, 2010b) identified individuals of any

> **BOX 1-2**
> **Statement of Task**
>
> An IOM committee will be convened to evaluate the results, study design, and methodological approaches that have been used to assess the relationship between sodium and health outcomes in literature published since the IOM Dietary Reference Intakes report on electrolytes, including relevant domestic and international literature. Of primary interest are the effects (potential benefits/adverse impacts) in the population generally and for population subgroups (particularly those with hypertension or prehypertension, persons 51 years of age and older, African Americans, and people with diabetes, chronic kidney disease, or congestive heart failure). The committee will prepare a report focusing on data after 2003, but include prior data in summary as needed on
>
> - the quality of the literature reviewed;
> - both the benefits and adverse outcomes of reduced population sodium intake, particularly to levels of 1,500 to less than 2,300 mg per day, and emphasizing relevant subgroups, including (a) people with hypertension or prehypertension, (b) people 51 years of age and older, (c) African Americans, (d) people with diabetes, (e) people with chronic kidney disease, and (f) people with congestive heart failure;
> - implications for population-based strategies to gradually reduce sodium intake; and
> - data and method gaps and suggested ways to address them.

age with hypertension, adults 51 years of age and older, and African Americans as at-risk subgroups within the general population. Specifically, the prevalence of hypertension is greater among African Americans than whites (Fields et al., 2004; Gillespie et al., 2011); this population group also is at greater risk of complications related to hypertension, such as stroke (Ayala et al., 2001; Giles et al., 1995) and kidney disease (Klag et al., 1996). Individuals who are 51 years of age and older and adults of any age who have prehypertension also are at greater risk of developing hypertension than are those who are younger and those with normal blood pressure (Gillespie et al., 2011; Lloyd-Jones et al., 2005; Vasan et al., 2001).

Finally, some evidence suggests that individuals with hypertension, diabetes, and CKD may be more sensitive to sodium than are nonaffected groups (Lifton et al., 2002) and, thus, are at greater risk of developing adverse outcomes such as CVD and stroke. Given the disproportionate level of risk of adverse health outcomes among these population subgroups, the committee identified for further evaluation evidence that evaluated sodium intake in relation to either benefit or risk for these groups and considered them as a subset apart from the overall U.S. population.

NHANES data indicate that these at-risk subgroups now comprise a majority of the U.S. population (HHS and USDA, 2010). Further, more than 90 percent of U.S. adults 50 years of age or older will develop hypertension in their lifetime (Vasan et al., 2002), and nearly 70 percent of men and 49 percent of women who develop hypertension in middle age will experience a CVD event by 85 years of age (Allen et al., 2012). Even though some population subgroups stand out because they are disproportionately affected by hypertension and its adverse health outcomes, they represent a large proportion of the general population. Thus, the committee considered that part of its task was to examine the effects of dietary sodium on health outcomes in the general population. Separately, the committee considered the data in these subgroups of special interest, both whether separate data were available on individuals in these subgroups, and whether the general population studies included analyses of specific subgroups.

THE STUDY PROCESS

The IOM established a committee of 12 members with expertise in nutrition, CVD, hypertension, diabetes, kidney disease, epidemiological studies, clinical trial design and data analysis, biostatistics, and evidence-based reviews. The committee met in closed session and by conference calls, and held a 2-day public workshop to gather information pertinent to the task.

To address its task, the committee first formulated a process to identify and weigh studies identified from its search of literature published from 2003 through 2012. The committee did not rate the studies for quality because the broad range of study designs and sodium intake assessment precluded the application of a uniform rating system. Instead, the papers were reviewed and assessed individually. The committee developed a strategy to qualitatively assess each relevant study and the totality of the evidence based on the variability of methodological approaches, study designs, the method to estimate sodium intake, confounder adjustment, and the number of relevant studies available. The committee's findings and conclusions are derived from its assessment of the evidence.

ORGANIZATION OF THE REPORT

This report reviews and evaluates the evidence on the potential for adverse health effects from reducing sodium intakes and discusses the findings, conclusions, and implications of the committee's assessment. The report is organized into five chapters. Chapter 1 describes the background

18

SODIUM INTAKE IN POPULATIONS

for the study and the statement of task. Chapter 2 describes in detail the literature search, including the process to identify relevant reports for review and the committee's methodological approach to assessing the quality of the evidence. The committee describes first the potential methodological issues in the studies reviewed. Then a detailed description of the literature search and process that the committee followed to review and assess the scientific literature is provided. Chapter 3 includes a summary of studies that examine the association between sodium intake and intermediate health outcomes, including blood pressure as a surrogate indicator of disease. The review includes an overview of studies that are representative of the current body of evidence. Chapter 4 describes the studies reviewed by the committee examining the association between sodium intake and direct health outcomes, identified from its search, and determined from its criteria as relevant for further review and assessment. Studies that did not meet the criteria for further consideration are summarized at the end of the chapter. Chapter 5 presents the committee's overarching findings, and the findings and conclusions for specific health outcomes, as well as the implications of its conclusions for reducing sodium intake in the population, and recommendations for future research.

Appendix A contains acronyms and abbreviations used throughout the report. Appendix B contains the committee members' biographical sketches. Appendix C presents the agendas of the open session and public workshop, respectively. Appendix D contains a depiction of direct and indirect pathways to adverse health effects related to sodium intake. Appendix E describes the literature search strategy. Appendix F contains tables with the summary of evidence for CVD, mortality, cancer, heart failure, metabolic syndrome, and diabetes as health outcomes.

REFERENCES

Allen, N., J. D. Berry, H. Ning, L. Van Horn, A. Dyer, and D. M. Lloyd-Jones. 2012. Impact of blood pressure and blood pressure change during middle age on the remaining lifetime risk for cardiovascular disease: The cardiovascular lifetime risk pooling project. *Circulation* 125(1):37-44.
ARS (Agricultural Research Service). 2010. *Sodium (mg): Usual intakes from food and water, 2003-2006, compared to adequate intakes and tolerable upper intake levels.* http://www.ars.usda.gov/SP2UserFiles/Place/12355000/pdf/0506/usual_nutrient_intake_sodium_2003-06.pdf (accessed March 18, 2013).
Ayala, C., K. J. Greenlund, J. B. Croft, N. L. Keenan, R. S. Donehoo, W. H. Giles, S. J. Kittner, and J. S. Marks. 2001. Racial/ethnic disparities in mortality by stroke subtype in the United States, 1995-1998. *American Journal of Epidemiology* 154(11):1057-1063.
Cogswell, M. E., Z. Zhang, A. L. Carriquiry, J. P. Gunn, E. V. Kuklina, S. H. Saydah, Q. Yang, and A. J. Moshfegh. 2012. Sodium and potassium intakes among US adults: NHANES 2003-2008. *American Journal of Clinical Nutrition* 96(3):647-657.

Fields, L. E., V. L. Burt, J. A. Cutler, J. Hughes, E. J. Roccella, and P. Sorlie. 2004. The burden of adult hypertension in the United States 1999 to 2000: A rising tide. *Hypertension* 44(4):398-404.

Giles, W. H., S. J. Kittner, J. R. Hebel, K. G. Losonczy, and R. W. Sherwin. 1995. Determinants of black-white differences in the risk of cerebral infarction: The National Health and Nutrition Examination Survey Epidemiologic Follow-up Study. *Archives of Internal Medicine* 155(12):1319-1324.

Gillespie, C., E. V. Kuklina, P. A. Briss, N. A. Blair, and Y. Hong. 2011. Vital signs: Prevalence, treatment, and control of hypertension—United States, 1999-2002 and 2005-2008. *Morbidity and Mortality Weekly Report* 60(4):103-108, http://www.cdc.gov/mmwr/preview/mmwrhtml/mm6004a4.htm (accessed April 10, 2013).

HHS and USDA (U.S. Department of Health and Human Services and U.S. Department of Agriculture). 2005. *The report of the Dietary Guidelines Advisory Committee on Dietary Guidelines for Americans, 2005.* Washington, DC: HHS. http://www.health.gov/dietaryguidelines/dga2005/default.htm#1 (accessed April 26, 2013).

HHS and USDA. 2010a. *Report of the Dietary Guidelines Advisory Committee on the Dietary Guidelines for Americans, 2010, to the Secretary of Agriculture and the Secretary of Health and Human Services.* Washington, DC: USDA/ARS. http://www.cnpp.usda.gov/Publications/DietaryGuidelines/2010/DGAC/Report/2010DGACReport-camera-ready-Jan11-11.pdf (accessed February 1, 2013).

HHS and USDA. 2010b. *Dietary Guidelines for Americans, 2010.* 7th ed. Washington, DC: U.S. Government Printing Office. http://www.cnpp.usda.gov/Publications/DietaryGuidelines/2010/PolicyDoc/PolicyDoc.pdf (accessed February 4, 2013).

IOM (Institute of Medicine). 2005. *Dietary reference intakes for water, potassium, sodium, chloride, and sulfate.* Washington, DC: The National Academies Press.

IOM. 2006. *Dietary reference intakes: The essential guide to nutrient requirements.* Washington, DC: The National Academies Press.

IOM. 2010. *Strategies to reduce sodium intake in the United States.* Washington, DC: The National Academies Press.

Klag, M. J., P. K. Whelton, B. L. Randall, J. D. Neaton, F. L. Brancati, C. E. Ford, N. B. Shulman, and J. Stamler. 1996. Blood pressure and end-stage renal disease in men. *New England Journal of Medicine* 334(1):13-18.

Lawes, C. M., S. V. Hoorn, and A. Rodgers. 2008. Global burden of blood-pressure-related disease, 2001. *The Lancet* 371(9623):1513-1518.

Lifton, R. P., F. H. Wilson, K. A. Choate, and D. S. Geller. 2002. Salt and blood pressure: New insight from human genetic studies. *Cold Spring Harbor Symposia on Quantitative Biology* 67:445-450.

Lloyd-Jones, D. M., J. C. Evans, and D. Levy. 2005. Hypertension in adults across the age spectrum: Current outcomes and control in the community. *Journal of the American Medical Association* 294(4):466-472.

Vasan, R. S., M. G. Larson, E. P. Leip, J. C. Evans, C. J. O'Donnell, W. B. Kannel, and D. Levy. 2001. Impact of high-normal blood pressure on the risk of cardiovascular disease. *New England Journal of Medicine* 345(18):1291-1297.

Vasan, R. S., A. Beiser, S. Seshadri, M. G. Larson, W. B. Kannel, R. B. D'Agostino, and D. Levy. 2002. Residual lifetime risk for developing hypertension in middle-aged women and men: The Framingham Heart Study. *Journal of the American Medical Association* 287(8):1003-1010.

2

Approach to Evidence Review

This chapter briefly summarizes the methodological challenges and the approach the committee took to assess the evidence on relationships between sodium intake and health outcomes. The chapter begins with a description of the approach used by the committee to identify relevant evidence for consideration in response to its task. The sources of evidence, questions that guided the literature search, the literature search strategy, and the process of selecting studies for detailed review are described, as are the criteria to critically appraise the individual studies. Finally, the chapter summarizes the advantages and limitations of different approaches used to estimate sodium intake, an important criterion that the committee used to assess the quality of the studies.

GENERAL APPROACH OF THE COMMITTEE

The committee focused its review and assessment on evidence for direct associations between sodium intake and risk of adverse health outcomes. Although intermediate markers serve as a means of tracking and predicting health status, the scientific community continues to debate their use. A recent Institute of Medicine (IOM) report, *Evaluation of Biomarkers and Surrogate Endpoints in Chronic Disease* (IOM, 2010), highlights the concerns about the use of intermediate markers and provides a framework to evaluate them. The drug development literature contains numerous examples, including with cardiovascular drugs, where treatment based on biomarkers did not correctly predict the ultimately studied clinical outcomes (e.g., Nissen and Wolski, 2007).

However, in the case of blood pressure, the recent report on biomarkers (IOM, 2010) points to its wide acceptance as a surrogate marker and to the recent evidence attributing 35 percent of myocardial infarction and stroke events, 49 percent of heart failure episodes, and 24 percent of premature deaths to high blood pressure (Lawes et al., 2008). Nevertheless, the committee recognizes that cardiovascular effects do not occur only as a result of blood pressure. For example, and as described in the 2010 IOM report, it is known that different classes of drugs can have multiple cardiovascular outcomes but not all of them are a result of their blood pressure lowering effects. For example, despite their effects on blood pressure, alpha blockers have been shown to have a higher risk of heart failure compared to diuretics (ALLHAT Collaborative Research Group, 2000). The committee also recognizes that, in addition to blood pressure effects, diets modified in sodium may execute their effects on health outcomes through other factors, including other dietary constituents. The committee did not address questions related to such potential additional mechanisms because such questions were not part of its statement of task.

The committee's assessment of the evidence reviewed was influenced by a number of factors. These included the variability in methodological approaches used to evaluate relationships between sodium intake and risk of health outcomes, study design, limitations in the quantitative measures of both dietary intake and urinary excretion of sodium, confounder adjustment, and the number of relevant studies available. Assessing the impact of sodium intake on health outcomes was further complicated by wide variability in intake ranges among studies. For example, in the studies reviewed, high sodium intake levels for examining associations with health outcomes ranged from about 2,700 to more than 10,000 mg per day. The lack of consistency between studies in defining sodium intakes at both high and low ends of the range of typical intakes among various population groups meant that the committee could not derive a numerical definition for high or low intakes in its findings and conclusions. Rather, it could consider sodium intake levels only within the context of an individual study. Thus, in its findings and conclusions, the committee's use of "high" or "low" sodium intake indicates levels in the ranges described in the evidence reviewed.

LITERATURE REVIEW

Sources of Information

The committee obtained data and information for its conclusions from several sources. The main source of information came from a review of the evidence in the scientific literature from 2003 through 2012. For this review, scientific literature searches were conducted by the study staff in consultation with National Academy of Sciences librarians and described below in detail and in Appendix E. The review was conducted by adhering closely to the recommendations of the 2011 IOM report *Finding What Works in Health Care: Standards for Systematic Reviews* (IOM, 2011). The committee also reviewed and considered summaries of the scientific information about the association of sodium dietary intake and direct health outcomes in recent reports from authoritative sources (HHS and USDA, 2005, 2010; IOM, 2005). Additional information was gathered during its public workshop (see Appendix C for workshop agenda). Invited presentations included a range of perspectives about relationships between sodium intake and health outcomes. In addition, stakeholders were invited to present their views on the topics presented. The committee considered unpublished data when provided by the public. Unpublished data, however, were not used as principal evidence for findings and conclusions; these data were used only as supportive evidence for committee's findings and conclusions, when appropriate. In a few circumstances, the committee concluded that additional analysis of the published data would add important information to a specific topic. In those cases, the authors of the study were contacted by staff and the analysis requested. As with other information, these additional analyses were included in the study public access file.

Questions

Based on the statement of task, the committee formulated the following questions to guide its literature search:

Question 1. What is the effect of reducing dietary sodium intake in all individuals compared to habitual intake on health outcomes (cardiovascular disease, heart failure, myocardial infarction, diabetes, mortality, stroke, bone disease, fractures, falls, headaches, kidney stones, skin reactions, immune function, thyroid disease, or cancer)?

Question 2. What is the effect of reducing dietary sodium intake in individuals with hypertension, prehypertension, those aged 51 years and older, African Americans, and individuals with diabetes, chronic kidney disease, or congestive heart failure, compared to habitual intake on health outcomes (cardiovascular disease, myocardial infarction, diabetes, mortality, stroke, bone disease, fractures, falls, headaches, kidney stones, skin reactions, immune function, thyroid disease, or cancer)?

Literature Search Strategy

Literature searches were conducted to identify studies to answer the committee's questions. The online databases used for these searches were the Cochrane Database of Systematic Reviews Embase, MedLine, PubMed, and Web of Science. A broad search was first performed to include all health outcomes. In addition, a number of searches targeted at specific outcomes identified by the committee were conducted. The specific outcomes were cardiovascular disease, congestive heart failure, hypertension, myocardial infarction, diabetes, mortality, stroke, bone disease, fractures, falls, headaches, kidney stones, chronic kidney disease, skin reactions, immune function, thyroid disease, and cancer. Table E-1 in Appendix E presents the search conducted in the MedLine database as an example of the searches conducted.

The searches were limited to peer-reviewed original research studies, systematic reviews, and meta-analyses published from January 1, 2003, through December 18, 2012, in the English language. Studies in all countries, of all sample sizes and of all follow-up periods were included. In addition, studies with all populations irrespective of health status, ages, races, and ethnicities were included. Case studies and case series as well as animal or in vitro studies were excluded in the search strategies. The studies included in relevant systematic reviews and meta-analyses were used only as background and to cross-check the references used so as to ensure the most complete review of the literature.

Selection of Studies

The abstracts of all studies identified by the targeted searches were reviewed by the IOM staff. A diagram illustrating the selection process is shown in Figure 2-1. Conference abstracts, nonsystematic reviews, public statements, policy studies, modeling studies, mechanistic studies, and studies where data on dietary sodium intake or a health outcome of interest were not collected were excluded. Studies with only intermediate outcomes were also excluded. Such studies, however, were summarized and are discussed in Chapter 3. The method used to estimate sodium intake was a key exclusion criterion. Only studies where a food frequency questionnaire (FFQ), a 24-hour recall, food record, or urinary excretion methods were used to estimate sodium assessment were selected for review. Among those, studies were excluded if the method to estimate sodium intake was not described in sufficient detail or for which numerical sodium levels were not calculated. In addition, studies that analyzed only the association between sodium/potassium ratio and a health outcome or that did not analyze the independent effect of sodium intake were excluded.

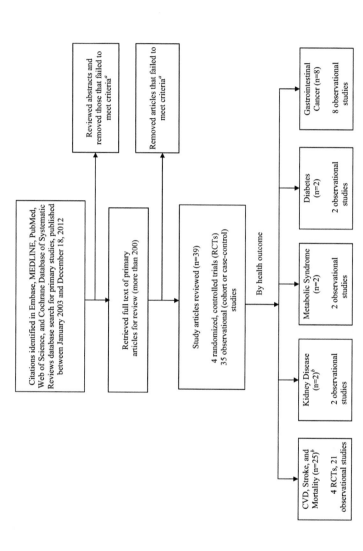

FIGURE 2-1 Flow diagram depicting the literature search strategy.

NOTE: CVD, cardiovascular disease; RCT, randomized controlled trial.

[a] See Chapter 2 for a description of the inclusion and exclusion criteria.

[b] One of the articles included in the Cardiovascular Disease, Stroke, and Mortality outcome group is also included in the Kidney Disease outcome group.

EVALUATION OF STUDIES

Review Process

Three committee members reviewed each of the studies individually and then as a group to deliberate and reach consensus on their strengths and weaknesses. Further, all studies were individually presented and discussed as a committee. The committee used summary tables to present details of the study designs (see evidence tables in Appendix F), the strengths and weaknesses based on the generalizability of the study population to the U.S. population and population subgroups of interest, and the methodological appropriateness of the studies reviewed (see Tables 4-1 through 4-8). The committee consulted the IOM report *Finding What Works in Health Care: Standards for Systematic Reviews* (IOM, 2011) to establish the overall review process, including determining the quality of the studies. The 2011 IOM report recommended that three elements be used to critically evaluate individual studies: methodological appropriateness (i.e., the risk of bias); relevance of the study's population, interventions, and outcomes measures; and the fidelity of implementation of interventions. The committee did not formally rate the studies for methodological quality because upon its review, it found the study designs were too varied and no generally accepted approach to performing such ratings for studies of dietary sodium existed. Instead, the studies were reviewed and assessed individually for their strengths and weaknesses. Evidence about associations between sodium intake and health outcomes was considered in its totality.

The committee also chose not to perform a formal meta-analysis of the available results. First, such a meta-analysis was deemed inappropriate for this review because of the wide differences in the methodologies of the studies, especially with respect to measurements of sodium intake and adjustment for confounders. Second, the studies were very different in the ranges of populations and sodium intakes examined, and the measures of dietary sodium could not readily be calibrated from study to study. Thus, aggregating them statistically would lead to a misleadingly simplistic summary statistic, rather than a meaningful synthesis. Instead, the committee evaluated each study in the context of its population, dietary sodium intake range, and methodological strengths and weaknesses in order to synthesize its findings and conclusions.

Criteria for Evaluating Evidence

Following identification of studies for review, the committee determined two broad criteria to critically appraise each study: generalizability to the populations of interest and methodological appropriateness (i.e., risk of bias).

Generalizability to Populations of Interest

The report *Evaluation of Biomarkers and Surrogate Endpoints in Chronic Disease* (IOM, 2010) recommends using relevance to the study population, interventions, and outcomes measures as one element to evaluate each study. In this evaluation, only relevant interventions and outcomes were selected for the final review. The relevance of the population under study to populations of interest in the statement of task (i.e., individuals with hypertension or prehypertension, those 51 years of age and older, African Americans, and individuals with diabetes, chronic kidney disease, and congestive heart failure) was used as an element of the evaluation.

Methodological Appropriateness

The body of evidence selected, based on the inclusion/exclusion criteria, included studies with wide diversity in methodological approaches. As discussed above, the committee did not rate studies for quality because, upon review, it found that the study designs, including randomized controlled trials (RCTs) and observational studies, were too varied. In addition, a consistent lack of calibration of approaches used to estimate sodium intake precluded a uniform rating system. Instead, each study was reviewed and critically appraised individually for its strengths and weaknesses (see Tables 4-1 through 4-8). Thus, the design and methodological approach of each study was of critical importance to the committee's evidence review.

The committee conducted a qualitative evaluation of the strengths and weaknesses of each eligible study based on the following criteria: size and characteristics of the study population, risk of bias, and fidelity of implementation of the intervention. The committee considered RCTs as the most valuable for determining the effect of sodium on health outcomes. Among observational studies, prospective cohort studies were considered the highest quality of design. Case-control studies were considered of lower quality as a study design, and the findings from these studies were given a lower relative weighting. Cross-sectional studies by themselves were included only as an indicator of a potential association between sodium intake and health outcome or to support results from other studies.

For RCTs, the committee identified specific criteria to assess risk of bias, including blinding, method of randomization, size and characteristics of the study population, drop-out rate, or relevance of the sodium intake level (intervention). The conformity of the method of implementation also was considered in appraising the quality of RCTs.

For observational studies, the committee evaluated risk of bias by evaluating strengths and weaknesses based on study design and length, method of measuring sodium intake as well as intake levels, and adjustment for confounders. To assess sodium intake measures, the committee used two broad categories: dietary intake and urinary excretion. Weaknesses in dietary intake measures included limitations in FFQs (e.g., few food items included, lack of validation with other sodium intake measurement methods, inaccurate food composition tables) and poor or missing validation of 24-hour food recalls.

Examples of weaknesses in urinary excretion measures included potential systematic errors in measuring sodium intake due to incomplete 24-hour urine collections, lack of reporting urine creatinine levels or body weight, and lack of validation of spot urine collection methods. Examples of weaknesses in adjustment of confounders included exclusion of appropriate confounders in the analytical model (e.g., age, gender, use of antihypertensive medications, caloric intake, dietary potassium intake, or other risk factors for the relevant health outcome) and inclusion of factors in the causal pathway of the effect as confounders in the analytical model. Interpretation of observational studies also took into account the possibility of reverse causality (i.e., the fact that an individual with lower sodium intake might experience an adverse health outcome, not because of sodium intake, but because of their health status, which in turn might lead to a low measured sodium intake).

METHODOLOGICAL ISSUES IN SODIUM RESEARCH: SODIUM INTAKE ASSESSMENT

The most common methods used to measure sodium intake are discussed below along with weaknesses and strengths. The descriptions are intended to illustrate that the appropriateness of the methodological approach depends on the application of the method and, most importantly, on the quality of its implementation. For example, although the 24-hour urine collection is portrayed as the "gold standard," it is important to assess the quality of its implementation.

Dietary Intake Assessment

Data on nutrient intakes among free-living individuals are collected most commonly by one or more of three different methods: 24-hour dietary recall, FFQ, or food record. Each method has its own advantages and disadvantages and, to be useful, must be selected to satisfy the research objectives. Factors that are considered in selecting a method to collect nutrient intake data include

- level of detail needed;
- research design/protocol;
- funding/resources (staff qualifications, resources for analyses);
- participant burden; and
- population characteristics (geography, ethnicity, education, health/ cognitive status).

24-Hour Dietary Recall

The 24-hour dietary recall is typically an interviewer-administered survey that asks respondents to recall everything they ate or drank in the previous 24 hours. In addition, the respondent is probed for such supplemental information as timing of meals, eating occasions, place where the meal is eaten, recipes used, and additions to foods. The 24-hour recall has the advantage that it can be used with most population groups because a high level of literacy is not required. Respondents are generally compliant because only a small amount of information is requested in the interview. This approach also does not affect eating behavior because it is retrospective. More accurate information about "usual" eating habits can be obtained when more than 1 day of recall data are collected (Conway et al., 2004).

A general limitation to the 24-hour recall is that respondents may underreport and/or underestimate foods or intake amounts. To minimize this problem, it is important that the recalls be collected by trained interviewers who follow a systematic protocol to elicit information. This requirement makes use of 24-hour recalls typically more costly than other diet assessment methods.

The National Health and Nutrition Examination Survey (NHANES) uses 24-hour recalls to assess dietary intake. Before 2003, a second dietary recall was collected for only a subset of respondents for quality assurance or to permit more accurate assessment of usual dietary intake. Beginning in 2001, a second recall was collected for a subset of respondents to permit more accurate assessment of usual dietary intake. Since 2003, two recalls have been collected for all respondents. The first recall is collected in person and the second recall is collected 3 to 10 days later by telephone.

Over the years, the methods used to estimate sodium intake in NHANES have been adjusted. As of 2005, NHANES surveys have included sodium consumed from tap and bottled water in the individual intake estimates. In earlier years, data on intake of tap and bottled water were collected, but the sodium contribution of water was not included in estimates of sodium intake. The methods used to capture contributions from salt used in cooking also have changed over time (Cogswell et al., 2012).

Food Frequency Questionnaire

The FFQ includes a finite list of foods and beverages and responses to indicate how often each item was consumed by the respondent over a specified period of time. FFQs are often used when a relatively low respondent burden is desired and/or when available resources preclude use of interviewer-administered methods. It has the advantage that it can be self-administered and, in some cases, responses can be optically scanned for ease of data entry and analysis. In addition, FFQs can capture usual intakes, rather than a single-day "snapshot" of intake. When applied to sodium intake analyses in U.S. populations, the FFQ generally does not include questions about salt added in cooking or at the table. Additionally, if a limited selection of foods is available to the respondent, the FFQ can also be limited in the detail and quantitative intakes it captures. Sodium levels in particular may differ widely by brand of processed foods, and these may be inadequately captured within the broad categories commonly used in these questionnaires. Overall, the FFQ serves best to rank order individuals rather than to estimate absolute levels of sodium intake (Boeing et al., 1997; Carithers et al., 2009; Fayet et al., 2011). Commonly used FFQs include the Block FFQ (Block et al., 1990), the Harvard FFQ (Feskanich et al., 1993; Oh et al., 2005), and the Diet History Questionnaire developed at the National Cancer Institute (Subar et al., 2001).

Food Records

A food record is a detailed description, usually in the form of a diary, of the types and amounts of foods, beverages, and/or supplements consumed over a specified time period, such as 3 days. Respondents are asked to record detailed information about the foods and beverages they consumed each day. The information requested can include food preparation methods, recipes, brand names of products, and portion sizes. Other information includes the type of meal consumed, and the time of day and location of consumption. Accurate responses require a high level of knowledge and motivation and may require training in how to record intakes accurately. The burden placed on respondents is often a disincentive to use food records and thus, compared to the FFQ and the 24-hour recall, it is infrequently used in intake studies. Another but important limitation is the potential for inaccuracy in recording actual intake of items due to respondents' awareness of or desire to exhibit behavior change (Freudenheim et al., 1987).

In summary, dietary sodium intake is a complex exposure, but it is possible to obtain useful information provided the right measures are used. FFQs, 24-hour recalls, and diet records each have measurement errors that must be taken into consideration in assessing and analyzing the data collected.

Use of Urine Samples to Estimate Dietary Sodium Intake

In addition to the 24-hour recall, FFQs, and diet records, another common method of assessing dietary sodium intake is through measuring urine sodium excretion. General considerations when analyzing urine samples for biochemical indicators of dietary intake, and particularly for sodium intake, are described in other publications (Bernstein and Willett, 2010). This section presents a brief comparison of the methods most commonly used in the papers that the committee reviewed (i.e., 24-hour urine collection, timed overnight urine collection, and spot urine analysis) as a demonstration of the challenges they present and factors that the committee considered when attempting to interpret and compare results among studies using different methods. In addition to the obvious sources of error, such as whether the individual is under pharmacological therapies that affect sodium excretion (e.g., diuretics when evaluating spot urine specimens), sources of systematic and random errors were considered when interpreting urine sodium data. The most common systematic errors in 24-hour urine specimens involve undercollection of specimens and the day specimens are collected (Dyer et al., 1994). In spot urine specimens, errors involve the substantial variations of the sodium concentration in the urine depending on the time of the collection (related to timing of meals) (Mann and Garber, 2010). When evaluating spot urine specimens and indexing to urine creatinine, differences in muscle mass are another important consideration, as this is a major determinant of urine creatinine concentration and may therefore influence the spot urine sodium-to-creatinine ratio (Heymsfield et al., 1983).

24-Hour Urine Sodium

As more than 90 percent of consumed sodium is absorbed and excreted in the urine (Holbrook et al., 1984; Pitts, 1974), an accurately collected 24-hour urine sodium excretion has been considered the clinical gold standard method to assess an individual's dietary sodium intake on the same day (Clark and Mossholder, 1986; Luft et al., 1982; McCullough et al., 1991; Schachter et al., 1980). A study by Luft and colleagues (1982) showed a correlation of 0.75 between actual sodium intake measured from the diet and average 24-hour urine sodium excretion over a 9-day collection. This method minimizes errors in sodium measurement due to changes in tonicity that occur throughout the day, and on different days. To estimate an individual's usual dietary intake, multiple urinary measurements of sodium are required over a 24-hour period (He et al., 1993; Liu et al., 1979, 1986, 1987). Collecting 24-hour urine specimens is cumbersome and challenging, and frequently inaccurate unless conducted in a monitored setting, such as an inpatient clinical research unit. In unmonitored settings, individuals may be less able to collect all of their urine due to physical limitations, urinary incontinence, neurological problems, or other health reasons. Thus, 24-hour urine sodium assessment is mostly available in populations with modest sample sizes. Outpatient 24-hour urine collections are available in some studies, but systematic undercollection is considered a major threat to validity. When accurately collected, 24-hour urine sodium measurements have particular advantages. They capture sodium from salt in food and salt added during the preparation of meals as well as at the table. They objectively measure dietary sodium intake, and they are independent of participant recall. On the other hand, systematic undercollection may introduce bias if those subjects with the greatest degree of comorbidity (and thus, highest risk of adverse events) are also those most likely to collect an insufficient amount of their urine. Undercollection will, of course, incorrectly classify individuals into the low-sodium categories.

Methods are available to assess and improve the accuracy of outpatient timed urine collections. Some studies average 24-hour urine sodium over multiple collections, which is likely to improve accuracy as the measure of usual sodium intake. Another method is to assess the 24-hour urine creatinine excretion. Because 24-hour urine creatinine excretion is not materially influenced by kidney function, and is tightly correlated to muscle mass, the total amount excreted over 24 hours should be relatively constant within an individual. Thus, undercollected 24-hour urine specimens may have lower-than-expected creatinine excretion. Conversely, overcollection will have higher-than-expected creatinine excretion. Prior studies suggest that usual 24-hour urine creatinine excretion is between 15-25 mg/kg/day in men and 10-20 mg/kg/day in women (Walser, 1987). Some studies have conducted sensitivity analyses that limited analysis to the subset with consistent 24-hour urine creatinine excretion measurements within individuals when multiple 24-hour urine collections were available. Others have excluded 24-hour urine specimens when 24-hour urine creatinine excretion measurements were implausibly low or high. Such maneuvers should decrease bias due to inaccurate collections and improve precision.

Alternative Urine Methods of Estimating Sodium Dietary Intake

Given the burden of collecting 24-hour urinary specimens and threats to validity due to collection errors, alternative methods of estimating sodium analysis have been developed, such as timed overnight (usually 8-hour) urine collections. Correlations between 24- and 8-hour overnight urine sodium are generally high. Various studies have shown good correlations (r=0.75 to 0.94) between overnight and 24-hour urine collections (He et al., 1993; Liu et al., 1979, 1986, 1987). Despite these high correlations, some authors have shown that the intra- and interindividual coefficients of variation were consistently greater for the 8-hour urine sodium than for the 24-hour urine sodium. Moreover, some studies suggest that individuals may excrete more sodium overnight than during the day, so 8-hour urine collections may introduce some systematic error (Dyer et al., 1987).

The ease of collecting spot urine sodium specimens provides an attractive third alternative. However, as specimens are collected at only one time point, they may introduce an important source of error because of temporal changes in urine tonicity as well as changes in urine sodium excretion that occur throughout the day and relative to the timing of meals. Use of spot specimens requires correction for urine tonicity, most commonly by indexing to urine creatinine (Godevithanage et al., 2010; Imai et al., 2011). A recent systematic review of studies comparing spot and 24-hour urine analysis for estimating sodium intake revealed substantial variations in the correlations (ranges=0.28-0.67) depending on the methods used (Ji et al., 2011). The authors highlighted the need for studies validating methods of sodium urine analysis as alternatives to the more cumbersome 24-hour urine sodium collection.

As discussed previously, urine creatinine excretion (a marker of collection accuracy) is influenced not only by urine tonicity, but also by muscle mass. Creatinine excretion is lower in persons with lower muscle mass. Thus, women, older individuals, those with lower body weight, and those of non-African descent have lower urine creatinine excretion, on average (Ix et al., 2011). A number of investigators have developed equations to estimate creatinine excretion on the basis of these demographic variables for various clinical indications (Cockcroft and Gault, 1976; Ix et al., 2011; Kawasaki et al., 1991). Kawasaki and colleagues (1982) developed one such equation. Subsequently, they multiplied spot urine sodium-to-creatinine ratios to the estimated creatinine excretion derived from their equation, in an attempt to improve the correlation of spot urine sodium-to-creatinine ratio with 24-hour urine sodium. They demonstrated that the correlation of this estimate to 24-hour urine sodium was 0.73. However, the equation to estimate creatinine excretion was derived in a Japanese population with a mean dietary sodium intake of approximately 5,000 mg per day, which is substantially higher than the typical sodium intake in the United States. O'Donnell and colleagues (2011) recently evaluated this equation in 105 participants in the Prospective Urban Rural Epidemiology study, a multinational study focused on noncommunicable disease epidemiology. In these individuals, the construct of spot urine sodium-to-creatinine ratio multiplied by the estimated creatinine excretion using the Kawasaki formula was correlated with measured 24-hour urine sodium excretion with a correlation coefficient of 0.55 (O'Donnell et al., 2011). This correlation is reasonably strong. Moreover, studies are consistent in demonstrating that correcting the spot urine sodium-to-creatinine ratio for creatinine excretion improves the correlation with measured 24-hour urine sodium compared to using spot urine sodium-to-creatinine ratio alone (Mann and Gerber, 2010). Further, despite reassuring correlation coefficients, the calibration of this technique in relation to actual dietary sodium intake is uncertain. The committee agreed that while the spot urine sodium-to-creatinine ratio multiplied by the estimated creatinine excretion construct appears useful to rank order persons with respect to their sodium intake, the absolute level of estimated dietary intake is uncertain in non-Asian populations.

In summary, multiple well-done 24-hour urine collections remain the gold standard method to assess dietary sodium intake, but bias due to inaccurate collection represents a major threat to validity. Moreover, most studies using 24-hour urine specimens do not provide data on urine volume, urine creatinine, and body weight, variables that would allow readers to evaluate whether 24-hour urine creatinine excretion is similar among sodium intake categories. Similar to survey methods such as the 24-hour diet recall, a single 24-hour, overnight 8-hour, or spot urine sodium measures are unlikely to capture habitual dietary sodium intake accurately. The correlation of spot urine sodium-to-creatinine ratio with 24-hour urine sodium is improved when adjusted for estimated creatinine excretion, which may provide a useful method to rank order participants based on dietary sodium intake in large epidemiologic studies. However, whether or not the estimated value of 24-hour urine sodium by this method is accurately calibrated to dietary sodium intake remains uncertain.

REFERENCES

ALLHAT Collaborative Research Group. 2000. Major cardiovascular events in hypertensive patients randomized to doxazosin vs chlorthalidone: The Antihypertensive and Lipid-Lowering Treatment to Prevent Heart Attack Trial (ALLHAT). *Journal of the American Medical Association* 283(15):1967-1975.

Bernstein, A. M., and W. C. Willett. 2010. Trends in 24-h urinary sodium excretion in the United States, 1957-2003: A systematic review. *American Journal of Clinical Nutrition* 92(5):1172-1180.

Block, G., M. Woods, A. Potosky, and C. Clifford. 1990. Validation of a self-administered diet history questionnaire using multiple diet records. *Journal of Clinical Epidemiology* 43(12):1327-1335.

Boeing, H., S. Bohlscheid-Thomas, S. Voss, S. Schneeweiss, and J. Wahrendorf. 1997. The relative validity of vitamin intakes derived from a food frequency questionnaire compared to 24-hour recalls and biological measurements: Results from the EPIC pilot study in Germany. *International Journal of Epidemiology* 26(Suppl 1):S82-S90.

Carithers, T. C., S. A. Talegawkar, M. L. Rowser, O. R. Henry, P. M. Dubbert, M. L. Bogle, H. A. Taylor Jr, and K. L. Tucker. 2009. Validity and calibration of food frequency questionnaires used with African-American adults in the Jackson Heart Study. *Journal of the American Dietetic Association* 109(7):1184-1193.e2.

Clark, A. J., and S. Mossholder. 1986. Sodium and potassium intake measurements: Dietary methodology problems. *American Journal of Clinical Nutrition* 43(3):470-476.

Cockcroft, D. W., and M. H. Gault. 1976. Prediction of creatinine clearance from serum creatinine. *Nephron* 16(1):31-41.

Cogswell, M. E., Z. Zhang, A. L. Carriquiry, J. P. Gunn, E. V. Kuklina, S. H. Sayday, Q. Yang, and A. L. Moshfegh. 2012. Sodium and potassium intakes among US adults: NHANES 2003-2008. *American Journal of Clinical Nutrition* 96(3):647-657.

Conway, J. M., L. A. Ingwersen, and A. J. Moshfegh. 2004. Accuracy of dietary recall using the USDA five-step multiple-pass method in men: An observational validation study. *Journal of the American Dietetic Association* 104(4):595-603.

Dyer, A. R., R. Stamler, R. Grimm, J. Stamler, R. Berman, F. C. Gosch, L. A. Emidy, P. Elmer, J. Fishman, N. Van Heel, and G. Civinelli. 1987. Do hypertensive patients have a different diurnal pattern of electrolyte excretion? *Hypertension* 10(4):417-424.

Dyer, A. R., M. Shipley, and P. Elliott. 1994. Urinary electrolyte excretion in 24-hours and blood pressure in the INTERSALT study. I. Estimates of reliability. *American Journal of Epidemiology* 139(9):927-939.

Fayet, F., V. Flood, P. Petocz, and S. Samman. 2011. Relative and biomarker-based validity of a food frequency questionnaire that measures the intakes of vitamin B_{12}, folate, iron, and zinc in young women. *Nutrition Research* 31(1):14-20.

Feskanich, D., E. B. Rimm, E. L. Giovannucci, G. A. Colditz, M. J. Stampfer, L. B. Litin, and W. C. Willett. 1993. Reproducibility and validity of food intake measurements from a semiquantitative food frequency questionnaire. *Journal of the American Dietetic Association* 93(7):790-796.

Freudenheim, J. L., N. E. Johnson, and R. L. Wardrop. 1987. Misclassification of nutrient intake of individuals and groups using one-, two-, three-, and seven-day food records. *American Journal of Epidemiology* 126(4):703-713.

Godevithanage, S., P. P. Kanankearachchi, M. P. Dissanayake, T. A. Jayalath, N. Chandrasiri, R. P. Jinasena, R. P. V. Kumarasiri, and C. D. A. Goonasekera. 2010. Spot urine osmolality/creatinine ratio in healthy humans. *Kidney and Blood Pressure Research* 33(4):291-296.

He, J., M. J. Klag, P. K. Whelton, J. Y. Chen, J. P. Mo, M. C. Qian, J. Coresh, P. S. Mo, and G. Q. He. 1993. Agreement between overnight and 24-hour urinary cation excretions in Southern Chinese men. *American Journal of Epidemiology* 137(11):1212-1220.

Heymsfield, S. B., C. Arteaga, C. M. McManus, J. Smith, and S. Moffitt. 1983. Measurement of muscle mass in humans: Validity of the 24-hour urinary creatinine method. *American Journal of Clinical Nutrition* 37(3):478-494.

HHS and USDA (U.S. Department of Health and Human Services and U.S. Department of Agriculture). 2005. *Dietary Guidelines for Americans, 2005.* 6th ed. Washington, DC: Government Printing Office. http://www.health.gov/dietaryguidelines/dga2005/document/pdf/DGA2005.pdf (accessed February 25, 2013).

HHS and USDA. 2010. *Report of the Dietary Guidelines Advisory Committee on the Dietary Guidelines for Americans, 2010, to the Secretary of Agriculture and the Secretary of Health and Human Services.* Washington, DC: USDA/ARS. http://www.cnpp.usda.gov/Publications/DietaryGuidelines/2010/DGAC/Report/2010DGACReport-camera-ready-Jan11-11.pdf (accessed February 1, 2013).

Holbrook, J. T., K. Y. Patterson, J. E. Bodner, L. W. Douglas, C. Veillon, J. L. Kelsay, W. Mertz, and J. C. Smith Jr. 1984. Sodium and potassium intake and balance in adults consuming self-selected diets. *American Journal of Clinical Nutrition* 40(4):786-793.

Imai, E., Y. Yasuda, M. Horio, K. Shibata, S. Kato, Y. Mizutani, J. Imai, M. Hayashi, H. Kamiya, Y. Oiso, T. Murohara, S. Maruyama, and S. Matsuo. 2011. Validation of the equations for estimating daily sodium excretion from spot urine in patients with chronic kidney disease. *Clinical and Experimental Nephrology* 15(6):861-867.

IOM (Institute of Medicine). 2005. *Dietary reference intakes for water, potassium, sodium, chloride, and sulfate.* Washington, DC: The National Academies Press.

IOM. 2010. *Evaluation of biomarkers and surrogate endpoints in chronic disease.* Washington, DC: The National Academies Press.

IOM. 2011. *Finding what works in health care: Standards for systematic reviews.* Washington, DC: The National Academies Press.

Ix, J. H., C. L. Wassel, L. A. Stevens, G. J. Beck, M. Froissart, G. Navis, R. Rodby, V. E. Torres, Y. Zhang, T. Greene, and A. S. Levey. 2011. Equations to estimate creatinine excretion rate: The CKD epidemiology collaboration. *Clinical Journal of the American Society of Nephrology* 6(1):184-191.

Ji, C., M. A. Miller, and F. P. Cappuccio. 2011. Comparisons of spot vs. 24-h urine samples for estimating salt intake. *Journal of Human Hypertension* 25:628.

Kawasaki, T., M. Ueno, K. Uezono, N. Kawazoe, S. Nakamuta, K. Ueda, and T. Omae. 1982. Average urinary excretion of sodium in 24-hours can be estimated from a spot-urine specimen. *Japanese Circulation Journal* 46(9):948-953.

Kawasaki, T., K. Uezono, K. Itoh, and M. Ueno. 1991. Prediction of 24-hour urinary creatinine excretion from age, body weight and height of an individual and its application. *Japanese Journal of Public Health* 38(8):567-574.

Lawes, C. M., S. V. Hoorn, and A. Rodgers. 2008. Global burden of blood-pressure-related disease, 2001. *The Lancet* 371(9623):1513-1518.

Liu, K., A. R. Dyer, R. S. Cooper, R. Stamler, and J. Stamler. 1979. Can overnight urine replace 24-hour urine collection to assess salt intake? *Hypertension* 1(5):529-536.

Liu, L., D. Zheng, and S. Lai. 1986. Variability in 24-hour urine sodium excretion in Chinese adults. *Chinese Medical Journal* 99(5):424-426.

Liu, L., Z. Deyu, and J. Lue. 1987. Variability of urinary sodium and potassium excretion in north Chinese men. *Journal of Hypertension* 5(3):331-335.

Luft, F. C., N. S. Fineberg, and R. S. Sloan. 1982. Overnight urine collections to estimate sodium intake. *Hypertension* 4(4):494-498.

Mann, S. J., and L. M. Gerber. 2010. Estimation of 24-h sodium excretion from a spot urine sample using chloride and creatinine dipsticks. *American Journal of Hypertension* 23(7):743-748.

McCullough, M. L., J. F. Swain, C. Malarick, and T. J. Moore. 1991. Feasibility of outpatient electrolyte balance studies. *Journal of the American College of Nutrition* 10(2):140-148.

Nissen, S. E., and K. Wolski. 2007. Effect of rosiglitazone on the risk of myocardial infarction and death from cardiovascular causes. *New England Journal of Medicine* 356(24):2457-2471.

O'Donnell, M. J., S. Yusuf, A. Mente, P. Gao, J. F. Mann, K. Teo, M. McQueen, P. Sleight, A. M. Sharma, A. Dans, J. Probstfield, and R. E. Schmieder. 2011. Urinary sodium and potassium excretion and risk of cardiovascular events. *Journal of the American Medical Association* 306(20):2229-2238.

Oh, K., F. B. Hu, E. Cho, K. M. Rexrode, M. J. Stampfer, J. E. Manson, S. Liu, and W. C. Willett. 2005. Carbohydrate intake, glycemic index, glycemic load, and dietary fiber in relation to risk of stroke in women. *American Journal of Epidemiology* 161(2):161-169.

Pitts, R. F. 1974. *Physiology of the kidney and body fluids.* 3rd ed. Chicago, IL: Year Book Medical Publishers, Inc.

Schachter, J., P. H. Harper, M. E. Radin, A. W. Caggiula, R. H. McDonald, and W. F. Diven. 1980. Comparison of sodium and potassium intake with excretion. *Hypertension* 2(5):695-699.

Subar, A. F., F. E. Thompson, V. Kipnis, D. Midthune, P. Hurwitz, S. McNutt, A. McIntosh, and S. Rosenfeld. 2001. Comparative validation of the Block, Willett, and National Cancer Institute Food Frequency Questionnaires: The Eating at America's Table Study. *American Journal of Epidemiology* 154(12):1089-1099.

Walser, M. 1987. Creatinine excretion as a measure of protein nutrition in adults of varying age. *Journal of Parenteral and Enteral Nutrition* 11(5 Suppl):73s-78s.

3

Sodium Intake and Intermediate Markers for Health Outcomes

BACKGROUND

In its statement of task, the committee was asked to review evidence for associations between dietary sodium and health outcomes published in the peer-reviewed literature since the last update of the report *Dietary Reference Intakes for Water, Potassium, Sodium, Chloride, and Sulfate* (IOM, 2005). This chapter first summarizes evidence and findings on sodium intake and intermediate markers for health outcomes reviewed in that report (IOM, 2005) and the 2010 report of the Dietary Guidelines Advisory Committee (DGAC) (HHS and USDA, 2010), then summarizes corresponding evidence published subsequently (from 2003 through 2012). This summary of new evidence is an overview of representative studies and includes only indicators or biomarkers of health outcomes and not clinical outcomes, which are reviewed in the next chapter. The evidence was considered by the committee to provide additional support for its findings and conclusions about associations between sodium intake and health outcomes; it was not used as a primary source of evidence.

Use of Biomarkers as Indicators of Health Outcomes

Biomarkers, as defined by the Biomarkers Definitions Working Group (Atkinson et al., 2001), are "a characteristic that is objectively measured and evaluated as an indicator of normal biological processes, pathogenic processes, or pharmacologic responses to a[n] . . . intervention." Applications of biomarkers include indicators of clinical endpoints (for example, in clinical trials) that denote how a study participant feels, functions, or survives; or in clinical practice, for example in risk stratification, disease prevention, screening, diagnosis, and monitoring. In public health practice, biomarkers serve as a means to track health status and for making recommendations for preventing, mitigating, and treating diseases or conditions at the population level (IOM, 2010). A related concept used in public health is the surrogate endpoint. The Institute of Medicine (IOM) Committee on Qualification of Biomarkers and Surrogate Endpoints in Chronic Disease defined surrogate endpoint as "a biomarker that is intended to substitute for a clinical endpoint. A surrogate endpoint is expected to predict clinical benefit (or harm, or lack of benefit or harm) based on epidemiologic, therapeutic, pathophysiologic, or other scientific evidence" (IOM, 2010, p. 23). This committee further identified blood pressure as "an exemplar surrogate endpoint for cardiovascular mortality and morbidity due to the levels and types of evidence that support its use" (IOM, 2010, p. 39).

Although biomarkers have wide utility in research, clinical practice, and public health, the biological complexity and variation among individuals is important to consider as potential sources of error in assessing the link between biomarkers and health outcomes. Further, and of critical importance, with the exception of blood pressure and HIV-1RNA (IOM, 2010), biomarkers predict clinical outcomes but are not necessarily directly correlated with them. Indeed, there are situations where treatment recommendations based on biomarkers have led to patient harm, once outcome studies were finally performed. For these reasons, the committee reviewed evidence from its literature search on a range of intermediate markers for health outcomes but did not include these studies in its assessment of relevant research in response to the task.

Given the limitations associated with most biomarkers as indicators of risk of adverse health outcome, as discussed in the Dietary Reference Intake (DRI) report (IOM, 2005), the DGAC (HHS and USDA, 2010) report, and the report on biomarkers and surrogate endpoints (IOM, 2010), blood pressure is widely recognized as a strong surrogate indicator for primary cardiovascular disease (CVD) clinical endpoints, such as myocardial infarction (MI) and stroke. The committee considered the strength of the evidence for blood pressure as a surrogate endpoint for risk of CVD and stroke and this evidence underpinned its assessment of new evidence on health outcomes. The committee summarizes new evidence for blood pressure as an indirect indicator of risk of CVD in this chapter, but does not include its assessment of this indicator in its comprehensive evidence review and analysis of sodium intake and direct health outcomes in Chapter 4.

BLOOD PRESSURE AS A BIOMARKER FOR
CARDIOVASCULAR DISEASE

Salt Sensitivity

Evidence presented to the committee in its data-gathering workshop (see Appendix C) and reviewed in IOM (2005) reinforces that reducing sodium intake can have widely varying effects among individuals. Nevertheless, the term "salt sensitive" has been used to describe those who experience the greatest reduction in blood pressure in response to decreased sodium intake. Conversely, "salt-resistant" individuals experience little change in blood pressure, even with dramatic changes in sodium intake (Weinberger, 1996). Interindividual heterogeneity in blood pressure in response to dietary sodium is described in IOM (2005, pp. 286-291), and includes findings from Obarzanek and colleagues (2003). This study examined blood pressure differences between two points: when sodium intake was similar and when sodium intake was decreased. A wide statistically normal distribution in measured blood pressure was seen among individuals at both intake levels. The standard deviation of the distribution change, however, was similar for both distributions, suggesting that the variability in blood pressure responses to reduced sodium intake likely occurred in response to factors unrelated to sodium intake. Biological variation in the physiological response to dietary sodium, mediated through the renin-angiotensin-aldosterone system (RAAS), has been postulated as a possible mechanism (Chamarthi et al., 2010).

Genetic Variation and Salt Sensitivity

Evidence examining a relationship between genetic variations and salt sensitivity and risk of high blood pressure suggests that such variations may be specific to certain population subgroups. Beeks et al. (2004) conducted a systematic review of reports across population groups on genetic factors associated with salt sensitivity. The review identified several candidate polymorphisms from among the studies reviewed. However, due to methodological differences, variations in the way salt sensitivity was defined, and a limited number of studies examining a given polymorphism, definitive conclusions could not be drawn. Other studies examining polymorphisms within specific population subgroups have identified genetic variants associated with racial/ethnic groups and risk of high blood pressure or hypertension, as illustrated in the following studies. These studies also show that multiple pathways are involved in blood pressure response to expression of gene variants.

Miyaki et al. (2005) used a food frequency questionnaire to estimate salt intake and genotyping for polymorphisms in the endothelial nitric oxide synthase (eNOS) gene, implicated in coronary artery disease and MI, to examine risk for hypertension in 281 healthy Japanese men. This study identified a specific mutation in the eNOS gene that, with a high-salt diet, was associated with a significant increase in blood pressure among affected men.

Zhang et al. (2010), in a small (n=329) cross-sectional study, identified the presence of cytochrome P450 3A polymorphisms in a group of Japanese adults. Blood pressure response to sodium intake, estimated by spot analysis of 24-hour urinary sodium excretion, was found to be associated with the frequency of expression of two allelic variants: a heterozygous modifier of blood pressure, and a homozygous variant in carriers that resulted in greater sensitivity to salt intake compared to noncarriers.

A small intervention study in 39 healthy adults in Sweden examined the influence of genetic variants in RAAS following a protocol of 4 weeks on a high-salt intake followed by 4 weeks on a low-salt intake (Dahlberg et al., 2007). Blood pressure measurements and 24-hour sodium excretions, taken at baseline and at the end of each dietary intervention, suggested enhanced salt sensitivity in normotensive participants carrying two variants of a gene associated with monogenic hypertension.

Kelly et al. (2009) and Gu et al. (2010) examined data from GenSalt, a large (n=1,906) 14-day intervention study carried out in rural China between 2003 and 2005, to identify gene variants that function in blood pressure regulation associated with salt sensitivity in a population consuming high levels of salt. Study participants consumed a low-sodium (3,000 mg per day) diet for the first 7 days, followed by a high-sodium (18,000 mg per day) diet for 7 days. Three timed urinary specimens were collected, one at baseline, then at the end of each intervention phase. Blood pressure measures were taken and genotyping for genetic polymorphisms was conducted for each participant. Kelly et al. (2009) identified two variants, one in the alpha-adducin gene and one in the guanine nucleotide binding protein beta polypeptide 3 genes. Gu et al. (2010) identified three novel variants from 11 RAAS candidate genes that were significantly associated with blood pressure response to salt intake. Another GenSalt study (Zhao et al., 2010) identified genetic variants in the angiotensin-converting enzyme 2 (ACE-2), a regulator of RAAS and the apelin receptor (a substrate of ACE-2), which were associated with blood pressure response to salt intake.

Together, these studies illustrate that a number of genetic variants are associated with salt sensitivity and susceptibility to high blood pressure associated with sodium intake, particularly among certain population subgroups. Additionally, individuals not seen to be at risk of hypertension may carry genetic polymorphisms that render them salt sensitive.

EVIDENCE ASSOCIATING SODIUM INTAKE
WITH BLOOD PRESSURE

Reports on Associations Between Sodium Intake and Blood Pressure

When it examined possible adverse effects of sodium overconsumption, the Panel on Dietary Reference Intakes for Electrolytes and Water (IOM, 2005) reviewed evidence on cardiovascular outcomes (stroke, coronary heart disease, left ventricular hypertrophy) and kidney disease and their associations with increased blood pressure. Evidence reviewed by the panel from meta-analyses of observational studies and clinical trials provided strong support for a link between high blood pressure and risk of CVD. Additional evidence from intervention studies consistently supported a dose-response relationship between dietary sodium intake and blood pressure in both normotensive and hypertensive individuals. Taken together, the panel concluded that reducing sodium intake therefore lowers blood pressure and thereby should decrease risk of CVD (IOM, 2005, pp. 351-357). However, evidence for other benefits associated with reduced sodium intake was inconclusive.

The 2010 DGAC (HHS and USDA, 2010) considered evidence published since the DRI report on water and electrolytes (IOM, 2005) in a systematic review on adverse effects of sodium on blood pressure, and included discussion related to sodium intake and risk of stroke, coronary heart disease, and kidney disease. This evidence along with previous evidence from the 2005 DGAC report (HHS and USDA, 2005) provided support for the committee's findings and recommendations for sodium intake in the general U.S. population. Although the 2010 DGAC report (HHS and USDA, 2010) also found variability in study design and intake assessment, and inconsistency in sodium measurements among the observational studies reviewed, collectively, the evidence was consistent with that identified in the previous DRI report on water and electrolytes (IOM, 2005) and showed a relationship between reducing sodium intake and lowering blood pressure.

In 2012, the World Health Organization (WHO) published its updated report *Guideline: Sodium Intake for Adults and Children* (WHO, 2012). The evidence base for the supporting literature review included epidemiological evidence, three systematic reviews (one of which was of randomized controlled trials [RCTs]) conducted by WHO, and a reanalysis of data from a fourth systematic review. The outcomes examined for associations with sodium intake were blood pressure in adults, all-cause mortality, CVD, stroke, and coronary heart disease in adults, potential adverse effects in adults, blood pressure in children, and potential adverse effects in children. The report found that the evidence for a relationship between sodium intake and blood pressure was of high quality, whereas that for sodium intake and all-cause mortality, CVD, stroke, and coronary heart disease was of lower quality. Nevertheless, collectively, the WHO concluded that the evidence reviewed supports the conclusion that any reduction in sodium intake is beneficial for most individuals regardless of initial sodium intake. The report further concluded that even a modest reduction in blood pressure from reducing sodium intake would have significant public health benefits.

New Evidence Associating Sodium Intake with Health Outcomes

Blood Pressure Response to Sodium Intake in Population Subgroups

A unique population identified in the INTERSALT study, the Yanomami Indians, exhibits consistently low systolic and diastolic blood pressures within the population and over their lifetimes (De Jesus Mancilha-Carvalho and De Souza e Silva, 2003). Members of this population maintain an active lifestyle and have a very low salt intake (assessed by 24-hour urinary sodium excretion) throughout life and have no indicators of risk of coronary heart disease. The finding that blood pressure in members of this population does not rise with age or stimulation of RAAS suggests the possibility of a relationship between salt intake and blood pressure response that can occur apart from physiological or genetic variation. Studies in more heterogeneous populations subjected to environmental factors not encountered by the Yanomami Indians show less consistency in blood pressure response to low sodium intakes.

At the other end of the spectrum are population subgroups that are at risk of hypertension, are considered prehypertensive, or have diagnosed hypertension. Evidence published since the report *Dietary Reference Intakes for Water, Potassium, Sodium, Chloride, and Sulfate* (IOM, 2005) includes additional subanalyses of data from the Dietary Approaches to Stop Hypertension (DASH)-Sodium trial as well as new evidence from the GenSalt intervention study in China and the Relationship between Hypertension and Salt Intake in Turkish Population (SALTURK) population-based epidemiological study in Turkey.

Bray et al. (2004) analyzed data collected in the DASH-Sodium trial to determine the effect of stepwise reduction in sodium intake on blood pressure as modified by hypertension status. The DASH-Sodium dietary pattern and a control diet representative of a typical American eating pattern and changes in blood pressure were compared within hypertensive versus nonhypertensive; non–African American versus African American; women versus men; 45 years of age and younger versus older than 45 years of age; and obese versus nonobese population groups at three sodium intake levels: 150 mmol (3,450 mg) (high), 100 mmol (2,300 mg) (intermediate), and 50 mmol (1,150 mg) (low) per day for 30 days. Changes in blood pressure were analyzed to determine an overall effect of sodium reduction, as well as the effects of differences between high versus low and between high versus intermediate and intermediate versus low sodium intake levels for each subgroup and for hypertensive status within each subgroup.

Analyses by subgroup found that, at each sodium level, the DASH diet significantly reduced systolic blood pressure among African Americans and women on low versus high, low versus intermediate, and intermediate versus high intake levels. Hypertensive participants significantly reduced blood pressure in a stepwise fashion at all sodium intake levels for both the control and DASH diets. Among nonhypertensive participants, significant reductions in systolic blood pressure were found at all levels for those on the control diet but only between the low versus high levels for those on the DASH diet. Analyses of hypertensive and nonhypertensive individuals within each subgroup found that, for the control diet, the association between stepwise sodium reduction and reduced blood pressure was statistically significant across all subgroups by hypertension status, except those in the 45 years of age or younger subgroup.

A subset of participants in the GenSalt study included 1,906 adults 18-60 years of age, who completed a 21-day dietary sodium and potassium intervention. The intervention included a low-salt (3,000 mg salt or ~1,200 mg sodium per day) diet for 7 days followed by a high-salt (18,000 mg salt or ~7,200 mg sodium per day) diet for 7 days with a 1,500 mg potassium supplement included with the high-salt diet. Analysis of data collected at baseline and during each intervention found that blood pressure and 24-hour urinary sodium excretion in about 75 percent of participants decreased during the low-salt (~1,200 mg sodium per day) intervention and increased during the high-salt (~7,200 mg sodium per day) intervention, providing support for previous observations of salt sensitivity across populations. This study further found that blood pressure response to sodium intake was more pronounced in women compared to men, and among older participants as well as those with higher baseline blood pressure levels, consistent with findings from the DASH trial (He et al., 2009a).

The SALTURK population-based epidemiological study examined associations between dietary salt intake and blood pressure response in 603 normotensive and 213 hypertensive adult men and women. Data on family history of hypertension, dietary habits, and daily salt consumption as well as other medical and demographic information were collected by in-person interview questionnaire. Blood pressure measurement and 24-hour urinary sodium excretion were collected on each participant. The mean intake of salt was estimated to be about 18,000 mg per day (~7,200 mg sodium per day) across the population. Within the population, salt intake was significantly higher among the obese, those living in rural areas, and those with lower education levels. In addition, men consumed more salt than women and salt intake increased significantly with increasing age. After adjusting for these factors, positive linear correlations were found between salt intake and systolic and diastolic blood pressure. More specifically, each 2,000 mg per day intake in salt (~800 mg sodium per day) correlated with an increase in systolic blood pressure of 5.8 mmHg.

Cook et al. (2005) examined blood pressure response to reduced dietary sodium intake in a 3-year prospective intervention as part of the Trials of Hypertension Prevention, Phase II (TOHP II) study. Participants with high-normal blood pressure ages 30-54 years with a body mass index (BMI) indicating overweight were randomized into either a sodium-reduction group (n=596) or a usual care control group (n=596). Those in the sodium-reduction group were counseled on how to reduce sodium intake to 3.5 g or less per day while those in the usual care group followed their usual diets. Both groups were assessed for mean blood pressure and urinary sodium excretion at baseline, 18, and 36 months. Analyses of data on the change in blood pressure corresponding with urinary sodium excretion showed a significant dose-response trend in decreasing systolic blood pressure that was strongest for participants who maintained a low sodium intake.

A meta-analysis of evidence from RCTs carried out between 1981 and 2004 found a modest but significant association between a reduction in salt intake and decreased blood pressure in children (He and MacGregor, 2006). More recently, He et al. (2008) analyzed data from the National Diet and Nutrition Survey to assess relationships between salt intake and blood pressure in a cross-sectional study of free-living children in the United Kingdom. The study population included a nationally representative sample of 1,658 children and adolescents 4-18 years of age. Dietary salt intake data were obtained from 7-day dietary records. Average intake of salt, excluding salt added in cooking or added at the table, ranged from 4,700 (±200) mg (~1,900 [±800] mg sodium) per day at 4 years of age up to 6,800 (±200) mg (~2,700 mg sodium) per day at 18 years of age. When analyzed by tertile of salt intake for each group 4-8 years of age, 9-13 years of age, and 14-18 years of age, there was a significant association between increasing salt intake and an increase in systolic blood pressure for those 4-8 and 9-13 years of age, but not for those 14-18 years of age. However, a positive association between salt intake and systolic blood pressure was seen when all age groups were analyzed together. No associations were found between salt intake and diastolic blood pressure or between salt use in cooking or at the table and systolic blood pressure, although there was a significant association with increasing pulse pressure compared to those who did not add salt in cooking or at the table.

Blood Pressure Response to Sodium Intake in
Normotensive Population Groups

Ducher et al. (2003) analyzed data from 296 normotensive young adults participating in a 2-year prospective study that collected blood pressure measurements, 24-hour urinary sodium, sodium-to-creatinine ratio, and dietary intake of sodium at the time of entry and exit from the study. At the end of the study period, a multiple regression analysis found, using a linear model, a significant association between both systolic and diastolic blood pressure and age, BMI, sodium intake, and alcohol intake (correlation coefficient for systolic blood pressure=0.37 and for diastolic blood pressure=0.47 [p<0.0001]). Considered as an independent variable, there was no significant association between blood pressure and sodium intake. To examine further for relationships between sodium intake and blood pressure within the study population, the investigators conducted a Z_{rho} analysis. This approach is based on an analysis of statistical dependence between values of the two variables (sodium intake and blood pressure) within individuals in the population, thereby allowing detection of a relationship between blood pressure and sodium intake among individuals in the study population when it was not significant for the population as a whole. Using this analytical approach, a significant correlation was found between sodium intake and diastolic blood pressure in 16 percent of the study population.

Another epidemiological study (Park et al., 2010) examined associations between dietary sodium, calcium, and potassium, and anthropometric measures of obesity in a subset of Korean adults (n=2,761) from the Korean National Health and Nutrition Examination Survey (KHANES III). Dietary sodium intake data were obtained from 24-hour dietary recall. This study, which included participants who were normotensive as well as prehypertensive and hypertensive, also found no correlation between sodium intake and either systolic or diastolic blood pressure, although an inverse correlation was found between calcium intake and blood pressure.

Several intervention studies in normotensive population groups examined blood pressure response to variations in dietary sodium intake but study findings were inconsistent. In a small (n=10) 6-week double-blind randomized crossover study, Starmans-Kool et al. (2011) examined the effect of changes in dietary salt intake on central blood pressure and wave reflection (a measure of dynamic blood flow) in healthy normotensive males, 22-40 years of age. Participants were normalized to 2,200 mg sodium per day the first week, then randomized into either 2,200 mg (128 mmol) sodium per day by capsule or matched placebo controls along with a diet containing 2,600-3,500 mg sodium per day. Data were collected daily on blood pressure, heart rate, arterial pressure, and blood flow velocity. The study found significantly elevated carotid systolic blood pressure but only small changes in brachial systolic blood pressure in participants receiving the high-sodium intervention in this population of young normotensive men.

Todd et al. (2012) used a 4-week single-blind randomized crossover trial to examine the effect of dietary sodium administered as tomato juice containing 0, 4,000, or 8,000 mg sodium per day on blood pressure and other measures of arterial function in 19 normotensive adults. All participants were normalized to a 2,600 mg sodium diet in a 2-week washout period between respective interventions. Blood pressure, urinary sodium, and other analyses were taken at baseline, 1, 2, and 4 weeks for each intervention. None of the interventions was found to be significantly associated with either systolic or diastolic blood pressure response despite an increase in urinary sodium corresponding with increased sodium intake.

To examine the influence of dietary sodium on nighttime blood pressure "dipping" in salt-sensitive compared to salt-resistant young adults (18-40 years of age, n=41) and children (8-15 years of age, n=28), Simonetti et al. (2010) placed participants on 7 days of a low-salt diet (300 mg [122 mg sodium] per day), followed by 7 days of the same diet with sodium chloride tablets (9,000 mg per day for adults and 120 mg per kg body weight per day for children) and measured 24-hour urine collections and oscillometric 24-hour ambulatory blood pressure during each test week. The low-salt diet was effective in reducing daytime systolic blood pressure among salt-sensitive but not salt-resistant adults and children. However, nighttime dipping in blood pressure was not significantly different between the two age groups, independent of salt sensitivity or salt intake.

A study of similar design carried out in an adult Amish population (n=465) obtained different results than those of Simonetti et al. (2010). This study by Montasser et al. (2011) also measured daytime and nighttime systolic blood pressure over two 6-day intervention periods. However, participants were subjected to a high-salt diet (6,440 mg per day) first, followed by a 6- to 14-day washout period, then a low-salt diet (980 mg [~380 mg sodium] per day). Ambulatory blood pressure was measured by a monitor worn by the participant on the last day of each intervention. In contrast to Simonetti et al. (2010), Montasser et al. (2011) found a significant reduction in systolic blood pressure response for both daytime and nighttime measures, particularly among women, older participants, and those with higher systolic blood pressure, following the low-salt intervention.

Daytime and nighttime systolic blood pressure response to high- and low-salt diet treatment also was studied in a group of normotensive adults from Sweden (Melander et al., 2007). The study used a randomized double-blind crossover design in which participants were provided a 3,000 mg salt (~1,200 mg sodium) per day baseline diet for 8 weeks. Then, in the crossover, each participant received a sodium chloride capsule (6,000 mg per day) with the baseline diet for 4 weeks and a placebo capsule for 4 weeks. Ambulatory systolic blood pressure measures over 24 hours and 24-hour urine collections were taken at baseline and at the end of each treatment period. Similar to Montasser et al. (2011), Melander et al. (2007) found that lowering salt intake by 6,000 mg per day significantly decreased systolic blood pressure for both daytime and nighttime measures.

A recent report (Coxson et al., 2013) used computer modeling of three different scenarios to evaluate the effect of sodium reduction over a period of 10 years on blood pressure and related health outcomes in the U.S. population. The report demonstrates sodium reductions, ranging from 4 to 40 percent, achieved a reduced risk of coronary heart disease, stroke, major CVD events, and all-cause mortality from all three scenarios. The greatest reduction in risk was associated with sodium reduction modeling based on the TOHP trials (discussed in Chapter 4).

Summary and Interpretation of Evidence

The studies reviewed by the committee, like those reviewed in IOM (2005) and DGAC (HHS and USDA, 2010), also show heterogeneity within and among population groups with regard to a relationship between sodium intake and blood pressure. The studies vary in the methodological approaches used to measure sodium intake, as well as in how they account for bias and potential confounding in their results. Nevertheless, considered collectively, the evidence in the studies reviewed here generally supports prior evidence that links excessive dietary sodium to elevated blood pressure in at-risk subgroups, particularly individuals with hypertension or prehypertension.

EVIDENCE ASSOCIATING SODIUM INTAKE WITH BIOMARKERS OF PROGRESSION OF PRIOR KIDNEY DISEASE

The progression of chronic kidney disease (CKD) appears to be related to dietary sodium intake, either through effects on blood pressure or other mechanisms. Examples of intermediate health outcomes that have addressed the effect of sodium intake in CKD are primarily those evaluating changes in urinary protein or albumin excretion (proteinuria). Studies evaluating the relationship of dietary sodium intake with risk of end-stage renal disease are addressed separately in Chapter 4.

Proteinuria

Although a reduced sodium intake is typically recommended to lower blood pressure in patients with CKD, the 2005 IOM report *Dietary Refer-*

ence Intakes for Water, Potassium, Sodium, Chloride and Sulfate found only one cross-sectional study associating sodium intake with albumin excretion at that time.

A systematic review of evidence on the relationship of dietary sodium markers for progression of CKD was published in 2006. The authors concluded that, on the basis of data on the effects of sodium intake on functional, structural, or pathological indicators such as glomerular filtration rate, image scanning, or proteinuria, modest dietary sodium chloride restriction for patients with CKD should be considered, especially for those with hypertension or proteinuria (Jones-Burton et al., 2006). The diversity in methodologies and poor quality of the studies was highlighted in this review.

More recently, other studies have emerged linking dietary sodium intake with intermediate markers of kidney disease, such as urinary proteinuria. Weir et al. (2012) conducted a large cohort study of kidney disease patients with and without diabetes to explore the relationship between dietary sodium (estimated from 24-hour sodium excretion) and proteinuria. In their regression model, urinary sodium alone explained 12 percent of the urinary protein variation and dietary potassium offset some of the increase in proteinuria.

In a small randomized controlled trial in African or African Caribbean hypertensive individuals, protein and protein-to-creatinine ratio excretion fell significantly with a reduced sodium intake diet of 5,000 mg salt (2,000 mg sodium) per day (Swift et al., 2005). This reduction seemed not to be related to a decrease in blood pressure but occurred concomitantly with a significant increase in the level of plasma renin activity (a measure of the activity of the RAAS).

Further, a 7-day trial with 43 Chinese hypertensive individuals showed that the sodium-restricted group (average of 96 mmol sodium per day measured by 24-hour urine excretion analysis) had significantly less urine protein when compared with the habitual diet group (average of 149 mmol sodium per day) (Yu et al., 2012). Another marker of CKD progression, urinary TGF-β-1, also decreased.

The importance of sodium in the management of CKD was highlighted in an observational study in which the authors correlated the intake of sodium to the use of antihypertensive medications (Boudville et al., 2005). The study suggests that in individuals with CKD and equivalent blood pressure control, higher sodium intakes (estimated by 24-hour urinary sodium analysis) are associated with greater use of antihypertensive medications. Effects on proteinuria also have been studied directly in patients treated for hypertension.

The relationship between 24-hour urinary sodium (one collection at home) and proteinuria was found again in a cross-sectional study conducted in Japan with individuals under treatment for hypertension (Ohta et al., 2012). In this study, the level of aldosterone, but not of renin activity, was correlated with variations in urinary sodium.

In a randomized crossover trial with 169 mildly hypertensive participants, reduction in salt intake estimated from levels in urine of 9,700 (high) to 6,500 (control) mg salt per day (3,880 and 2,600 mg of sodium per day for high and control, respectively) resulted in significant reductions in albumin-to-creatinine ratios after 6-week intervention (He et al., 2009b).

Other studies have been designed to explore the interplay of drug therapies with dietary sodium (or salt) intake. For example, the effects on proteinuria of a dietary restriction or combination of an angiotensin receptor blocker and an angiotensin-converting enzyme inhibitor was tested in a small randomized controlled study with patients with nephropathy but not diabetes in the Netherlands (Slagman et al., 2011). The results indicated that a reduced-sodium diet (1,200 versus 4,800 mg per day for 6 weeks) was more effective than the angiotensin receptor blocker in reducing proteinuria in these patients. These results agree with those from other studies (Vogt et al., 2008).

Another small randomized controlled study also indicated that a lower-sodium diet had significant effects on decreasing proteinuria, regardless of the therapeutic intervention with an angiotensin II receptor antagonist, a diuretic, or both (Waanders et al., 2009).

Renin-Angiotensin-Aldosterone System

The RAAS is related to the progression of kidney disease. It also has been found to be associated with changes in sodium intake (Abiko et al., 2009; Alderman et al., 1991). Among the elements of the RAAS, plasma renin levels or activity have received substantial attention. The IOM report *Dietary Reference Intakes for Water, Potassium, Sodium, Chloride, and Sulfate* (IOM, 2005) suggested that the inverse relationship between renin and sodium intake appears to be curvilinear and it occurs at levels less than 2,300 mg per day. Below that level, and especially below 1,000 mg per day, renin rises exponentially.

Plasma renin activity (PRA) also has been proposed as a predictor of cardiovascular risk. Therefore, increases in PRA due to low sodium intake could have adverse health effects in the population. In a large, stable population of high-risk patients with atherosclerosis and/or diabetes, for example, PRA was an independent predictor of major vascular events and mortality (Verma et al., 2011). PRA also has been related to cardiovascular risk factors (e.g., hypertension, left ventricular hypertrophy, lipid levels) and with insulin resistance.

A recent review, however, questions the validity of plasma renin as a biomarker for cardiovascular events. The authors concluded that even though most studies have shown a positive association, conclusions are difficult to make based on differences across the studies (Volpe et al., 2012).

In the Antihypertensive Lipid-Lowering Treatment to Prevent Heart Attack Trial, participants assigned to diuretics, which increase RAAS, had similar CVD event risk and lower risk of heart failure compared to those on calcium channel blockers (Furberg et al., 2002). A recent Cochrane review update that included 167 RCTs from 1950 to 2011 addressed the effect of sodium intake on blood pressure, renin, aldosterone, catecholamines, cholesterol, and triglyceride in healthy persons with high or normal blood pressure (Graudal, 2012). The sodium intake, in the range of 120 to 150 mmol (2,760 to 3,450 mg) per day in three studies, and less than 120 mmol (2,760 mg) per day in all other studies, was estimated from 24-hour urine collection or from a sample of a minimum of 8 hours. This update highlighted important results, such as significant increases in urine aldosterone and blood renin that were proportional to estimated sodium intake. These results are in agreement with prior meta-analysis of trials (He and MacGregor, 2002; Jürgens and Graudal, 2003). However, experts still do not agree about the significance for health outcomes of the increases of blood renin levels with lower sodium intake.

REFERENCES

Abiko, H., T. Konta, Z. Hao, S. Takasaki, K. Suzuki, K. Ichikawa, A. Ikeda, Y. Shibata, Y. Takeishi, S. Kawata, T. Kato, and I. Kubota. 2009. Factors correlated with plasma renin activity in general Japanese population. *Clinical and Experimental Nephrology* 13(2):130-137.

Alderman, M. H., S. Madhavan, W. L. Ooi, H. Cohen, J. E. Sealey, and J. H. Laragh. 1991. Association of the renin-sodium profile with the risk of myocardial infarction in patients with hypertension. *New England Journal of Medicine* 324(16):1098-1104.

Atkinson, A. J., Jr., W. A. Colburn, V. G. DeGruttola, D. L. DeMets, G. J. Downing, D. F. Hoth, J. A. Oates, C. C. Peck, R. T. Schooley, B. A. Spilker, J. Woodcock, and S. L. Zeger. 2001. Biomarkers and surrogate endpoints: Preferred definitions and conceptual framework. *Clinical Pharmacology and Therapeutics* 69(3):89-95.

Beeks, E., A. G. H. Kessels, A. A. Kroon, M. M. Van Der Klauw, and P. W. De Leeuw. 2004. Genetic predisposition to salt-sensitivity: A systematic review. *Journal of Hypertension* 22(7):1243-1249.

Boudville, N., S. Ward, M. Benaroia, and A. A. House. 2005. Increased sodium intake correlates with greater use of antihypertensive agents by subjects with chronic kidney disease. *American Journal of Hypertension* 18(10):1300-1305.

Bray, G. A., W. M. Vollmer, F. M. Sacks, E. Obarzanek, L. P. Svetkey, L. J. Appel, and DASH Collaborative Research Group. 2004. A further subgroup analysis of the effects of the DASH diet and three dietary sodium levels on blood pressure: Results of the DASH-Sodium Trial. [Erratum appears in *American Journal of Cardiology* 2010 Feb 15;105(4):579]. *American Journal of Cardiology* 94(2):222-227.

Chamarthi, B., J. S. Williams, and G. H. Williams. 2010. A mechanism for salt-sensitive hypertension: Abnormal dietary sodium-mediated vascular response to angiotensin-II. *Journal of Hypertension* 28(5):1020-1026.

Cook, N. R., S. K. Kumanyika, J. A. Cutler, and P. K. Whelton. 2005. Dose-response of sodium excretion and blood pressure change among overweight, nonhypertensive adults in a 3-year dietary intervention study. *Journal of Human Hypertension* 19(1):47-54.

Coxson, P. G., N. R. Cook, M. Joffres, Y. Hong, D. Orenstein, S. M. Schmidt, and K. Bibbins-Domingo. 2013. Mortality benefits from US population-wide reduction in sodium consumption: Projections from 3 modeling approaches. *Hypertension* 61(3):564-570.

Dahlberg, J., L. O. Nilsson, F. von Wowern, and O. Melander. 2007. Polymorphism in NEDD4L is associated with increased salt sensitivity, reduced levels of P-renin and increased levels of Nt-proANP. *PLoS ONE [Electronic Resource]* 2(5):e432.

De Jesus Mancilha-Carvalho, J., and N. A. De Souza e Silva. 2003. The Yanomami Indians in the INTERSALT Study. *Os Yanomami no INTERSALT* 80(3):289-300.

Ducher, M., J. P. Fauvel, M. Maurin, M. Laville, P. Maire, C. Z. Paultre, and C. Cerutti. 2003. Sodium intake and blood pressure in healthy individuals. *Journal of Hypertension* 21(2):289-294.

Furberg, C. D., J. T. Wright Jr, B. R. Davis, J. A. Cutler, M. Alderman, H. Black, W. Cushman, R. Grimm, L. J. Haywood, F. Leenen, S. Oparil, J. Probstfield, P. Whelton, C. Nwachuku, D. Gordon, M. Proschan, P. Einhom, C. E. Ford, L. B. Piller, I. K. Dunn, D. Goff, S. Pressel, J. Bettencourt, B. DeLeon, L. M. Simpson, J. Blanton, T. Geraci, S. M. Walsh, C. Nelson, M. Rahman, A. Juratovac, R. Pospisil, L. Carroll, S. Sullivan, J. Russo, G. Barone, R. Christian, S. Feldman, T. Lucente, D. Calhoun, K. Jenkins, P. McDowell, J. Johnson, C. Kingry, J. Alzate, K. L. Margolis, L. A. Holland-Klemme, B. Jaeger, J. Williamson, G. Louis, P. Ragusa, A. Williard, R. L. S. Ferguson, J. Tanner, J. Eckfeldt, R. Crow, and J. Pelosi. 2002. Major outcomes in high-risk hypertensive patients randomized to angiotensin-converting enzyme inhibitor or calcium channel blocker vs diuretic: The Antihypertensive and Lipid-Lowering Treatment to Prevent Heart Attack Trial (ALLHAT). *Journal of the American Medical Association* 288(23):2981-2997.

Graudal, N. A., T. Hubeck-Graudal, and G. Jürgens. 2012. Effects of low-sodium diet vs. high-sodium diet on blood pressure, renin, aldosterone, catecholamines, cholesterol, and triglyceride (Cochrane Review). *American Journal of Hypertension* 25(1):1-15.

Gu, D., T. N. Kelly, J. E. Hixson, J. Chen, D. Liu, J. C. Chen, D. C. Rao, J. Mu, J. Ma, C. E. Jaquish, T. K. Rice, C. Gu, L. L. Hamm, P. K. Whelton, and J. He. 2010. Genetic variants in the renin-angiotensin-aldosterone system and salt sensitivity of blood pressure. *Journal of Hypertension* 28(6):1210-1220.

He, F. J., and G. A. MacGregor. 2002. Effect of modest salt reduction on blood pressure: A meta-analysis of randomized trials. Implications for public health. *Journal of Human Hypertension* 16(11):761-770.

He, F. J., and G. A. MacGregor. 2006. Importance of salt in determining blood pressure in children: Meta-analysis of controlled trials. *Hypertension* 48(5):861-869.

He, F. J., N. M. Marrero, and G. A. MacGregor. 2008. Salt and blood pressure in children and adolescents. *Journal of Human Hypertension* 22(1):4-11.

He, J., D. Gu, J. Chen, C. E. Jaquish, D. C. Rao, J. E. Hixson, J. C. Chen, X. Duan, J. F. Huang, C. S. Chen, T. N. Kelly, L. A. Bazzano, and P. K. Whelton. 2009a. Gender difference in blood pressure responses to dietary sodium intervention in the GenSalt study. *Journal of Hypertension* 27(1):48-54.

He, F. J., M. Marciniak, E. Visagie, N. D. Markandu, V. Anand, R. N. Dalton, and G. A. MacGregor. 2009b. Effect of modest salt reduction on blood pressure, urinary albumin, and pulse wave velocity in white, black, and Asian mild hypertensives. *Hypertension* 54(3):482-488.

HHS and USDA (U.S. Department of Health and Human Services and U.S. Department of Agriculture). 2005. *2005 Report of the Dietary Guidelines Advisory Committee.* Washington, DC: HHS. http://www.health.gov/dietaryguidelines/dga2005/report (accessed February 25, 2013).

HHS and USDA. 2010. *Report of the Dietary Guidelines Advisory Committee on the Dietary Guidelines for Americans, 2010, to the Secretary of Agriculture and the Secretary of Health and Human Services.* Washington, DC: USDA/ARS. http://www.cnpp.usda.gov/Publications/DietaryGuidelines/2010/DGAC/Report/2010DGACReport-camera-ready-Jan11-11.pdf (accessed February 1, 2013).

IOM (Institute of Medicine). 2005. *Dietary reference intakes for water, potassium, sodium, chloride, and sulfate.* Washington, DC: The National Academies Press.

IOM. 2010. *Evaluation of biomarkers and surrogate endpoints in chronic disease.* Washington, DC: The National Academies Press.

Jones-Burton, C., S. I. Mishra, J. C. Fink, J. Brown, W. Gossa, G. L. Bakris, and M. R. Weir. 2006. An in-depth review of the evidence linking dietary salt intake and progression of chronic kidney disease. *American Journal of Nephrology* 26(3):268-275.

Jürgens, G., and N. A. Graudal. 2003. Effects of low sodium diet versus high sodium diet on blood pressure, renin, aldosterone, catecholamines, cholesterols, and triglyceride. *Cochrane Database of Systematic Reviews* (1).

Kelly, T. N., T. K. Rice, D. Gu, J. E. Hixson, J. Chen, D. Liu, C. E. Jaquish, L. A. Bazzano, D. Hu, J. Ma, C. C. Gu, J. Huang, L. L. Hamm, and J. He. 2009. Novel genetic variants in the alpha-adducin and guanine nucleotide binding protein beta-polypeptide 3 genes and salt sensitivity of blood pressure. *American Journal of Hypertension* 22(9):985-992.

Melander, O., F. V. Wowern, E. Frandsen, P. Burri, G. Willsteen, M. Aurell, and U. L. Hulthen. 2007. Moderate salt restriction effectively lowers blood pressure and degree of salt sensitivity is related to baseline concentration of renin and N-terminal atrial natriuretic peptide in plasma. *Journal of Hypertension* 25(3):619-627.

Miyaki, K., S. Tohyama, M. Murata, H. Kikuchi, I. Takei, K. Watanabe, and K. Omae. 2005. Salt intake affects the relation between hypertension and the T-786C polymorphism in the endothelial nitric oxide synthase gene. *American Journal of Hypertension* 18(12 Pt 1):1556-1562.

Montasser, M. E., J. A. Douglas, M. H. Roy-Gagnon, C. V. Van Hout, M. R. Weir, R. Vogel, A. Parsa, N. I. Steinle, S. Snitker, N. H. Brereton, Y. P. Chang, A. R. Shuldiner, and B. D. Mitchell. 2011. Determinants of blood pressure response to low-salt intake in a healthy adult population. *Journal of Clinical Hypertension* 13(11):795-800.

Obarzanek, E., M. A. Proschan, W. M. Vollmer, T. J. Moore, F. M. Sacks, L. J. Appel, L. P. Svetkey, M. M. Most-Windhauser, and J. A. Cutler. 2003. Individual blood pressure responses to changes in salt intake: Results from the DASH-sodium trial. *Hypertension* 42(4 I):459-467.

Ohta, Y., T. Tsuchihashi, and K. Kiyohara. 2012. Influence of salt intake on target organ damages in treated hypertensive patients. *Clinical and Experimental Hypertension* 34(5):316-320.

Park, J., J. S. Lee, and J. Kim. 2010. Relationship between dietary sodium, potassium, and calcium, anthropometric indexes, and blood pressure in young and middle aged Korean adults. *Nutrition Research and Practice* 4(2):155-162.

Simonetti, G. D., S. Farese, F. Aregger, D. Uehlinger, F. J. Frey, and M. G. Mohaupt. 2010. Nocturnal dipping behaviour in normotensive white children and young adults in response to changes in salt intake. *Journal of Hypertension* 28(5):1027-1033.

Slagman, M. C. J., F. Waanders, M. H. Hemmelder, A. J. Woittiez, W. M. T. Janssen, H. J. Lambers Heerspink, G. Navis, and G. D. Laverman. 2011. Moderate dietary sodium restriction added to angiotensin converting enzyme inhibition compared with dual blockade in lowering proteinuria and blood pressure: Randomised controlled trial. *British Medical Journal (Online)* 343(7818).

Starmans-Kool, M. J., A. V. Stanton, Y. Y. Xu, S. A. McG. Thom, K. H. Parker, and A. D. Hughes. 2011. High dietary salt intake increases carotid blood pressure and wave reflection in normotensive healthy young men. *Journal of Applied Physiology* 110(2):468-471.

Swift, P. A., N. D. Markandu, G. A. Sagnella, F. J. He, and G. A. MacGregor. 2005. Modest salt reduction reduces blood pressure and urine protein excretion in black hypertensives: A randomized control trial. *Hypertension* 46(2):308-312.

Todd, A. S., R. J. MacGinley, J. B. W. Schollum, S. M. Williams, W. H. F. Sutherland, J. I. Mann, and R. J. Walker. 2012. Dietary sodium loading in normotensive healthy volunteers does not increase arterial vascular reactivity or blood pressure. *Nephrology* 17(3):249-256.

Verma, S., M. Gupta, D. T. Holmes, L. Xu, H. Teoh, S. Gupta, S. Yusuf, and E. M. Lonn. 2011. Plasma renin activity predicts cardiovascular mortality in the Heart Outcomes Prevention Evaluation (HOPE) study. *European Heart Journal* 32(17):2135-2142.

Vogt, L., F. Waanders, F. Boomsma, D. de Zeeuw, and G. Navis. 2008. Effects of dietary sodium and hydrochlorothiazide on the antiproteinuric efficacy of losartan. *Journal of the American Society of Nephrology* 19(5):999-1007.

Volpe, M., A. Battistoni, D. Chin, S. Rubattu, and G. Tocci. 2012. Renin as a biomarker of cardiovascular disease in clinical practice. *Nutrition, Metabolism and Cardiovascular Diseases* 22(4):312-317.

Waanders, F., V. S. Vaidya, H. van Goor, H. Leuvenink, K. Damman, I. Hamming, J. V. Bonventre, L. Vogt, and G. Navis. 2009. Effect of renin-angiotensin-aldosterone system inhibition, dietary sodium restriction, and/or diuretics on urinary kidney injury molecule 1 excretion in nondiabetic proteinuric kidney disease: A post hoc analysis of a randomized controlled trial. *American Journal of Kidney Diseases* 53(1):16-25.

Weinberger, M. H. 1996. Salt sensitivity of blood pressure in humans. *Hypertension* 27(3 II):481-490.

Weir, M. R., R. R. Townsend, J. C. Fink, V. Teal, S. M. Sozio, C. A. Anderson, L. J. Appel, S. Turban, J. Chen, J. He, N. Litbarg, A. Ojo, M. Rahman, L. Rosen, S. Steigerwalt, L. Strauss, and M. M. Joffe. 2012. Urinary sodium is a potent correlate of proteinuria: Lessons from the Chronic Renal Insufficiency Cohort Study. *American Journal of Nephrology* 36(5):397-404.

WHO (World Health Organization). 2012. *Guideline: Sodium intake for adults and children.* Geneva: WHO. http://www.who.int/nutrition/publications/guidelines/sodium_intake_printversion.pdf (accessed February 25, 2013).

Yu, W., S. Luying, W. Haiyan, and L. Xiaomei. 2012. Importance and benefits of dietary sodium restriction in the management of chronic kidney disease patients: Experience from a single Chinese center. *International Urology and Nephrology* 44(2):549-556.

Zhang, L., K. Miyaki, W. Wang, and M. Muramatsu. 2010. CYP3A5 polymorphism and sensitivity of blood pressure to dietary salt in Japanese men. *Journal of Human Hypertension* 24(5):345-350.

Zhao, Q., D. Gu, T. N. Kelly, J. E. Hixson, D. C. Rao, C. E. Jaquish, J. Chen, J. Huang, C. S. Chen, C. C. Gu, P. K. Whelton, and J. He. 2010. Association of genetic variants in the apelin-APJ system and ACE2 with blood pressure responses to potassium supplementation: The GenSalt study. *American Journal of Hypertension* 23(6):606-613.

4

Sodium Intake and Health Outcomes

This chapter reviews and assesses new evidence for associations between dietary sodium intake and outcomes published in the peer-reviewed literature through 2012. The health outcomes reviewed by the committee include cardiovascular disease (CVD), including stroke CVD mortality and all-cause mortality, congestive heart failure (CHF), chronic kidney disease (CKD), diabetes, cancer, and "other" outcomes, such as asthma and depression.

An estimated 76.4 million adults 20 years of age and older in the United States have high blood pressure (Roger et al., 2011). Mean dietary intake of sodium among the general U.S. population averages 3,400 mg daily, while federal nutrition policy guidance, the *Dietary Guidelines for Americans 2010* (HHS and USDA, 2010a), recommends sodium intakes of less than 2,300 mg daily for adolescents and adults 14 years of age and older, and 1,500 mg daily for African Americans, individuals 51 years of age and older, and individuals with hypertension, diabetes, or CKD. Evidence underlying this recommendation can be found in a number of sources, including the report *Dietary Reference Intakes for Water, Sodium, Chloride, and Sulfate* (IOM, 2005), and the *Report of the Dietary Guidelines Advisory Committee on the Dietary Guidelines for Americans, 2010* (DGAC) (HHS and USDA, 2010b). Excess dietary sodium has been identified as a potential etiologic risk factor for CVD, based on evidence for a dose-dependent increase in blood pressure in response to increasing sodium intake, as well as evidence from studies published before 2003 of sodium intake and risk of stroke or coronary heart disease (IOM, 2005).

The DGAC report (HHS and USDA, 2010b) included a review of evidence on the impact of dietary patterns low in sodium and/or low in saturated fat and high in potassium (e.g., DASH [Dietary Approaches to Stop Hypertension] and Mediterranean diets) on risk of CVD, stroke, and mortality and concluded that plant-based, lower-sodium dietary patterns had a beneficial impact on CVD risk. However, the dietary patterns included in the evidence review included dietary modifications other than sodium reduction that have been shown to have a cumulative impact on risk of CVD and related diseases. These include increased potassium, reduced intake of saturated and *trans* fats, and increased intake of dietary fiber. Based on their review of the evidence, the DGAC (HHS and USDA, 2010b) concluded that reduced risk of CVD, stroke, and disease-related mortality was associated with total dietary modification. Nevertheless, decreasing sodium intake has a potential role in reducing risk of CVD, stroke, and mortality, and this was the primary focus of the committee's review.

CARDIOVASCULAR DISEASE, STROKE, AND MORTALITY

As noted in Chapter 3, blood pressure is used as a surrogate indicator for CVD, stroke, and mortality risk, especially among individuals who are already at risk of disease. The committee's review of the strength of new evidence for dietary sodium and its effects on blood pressure concurred with previously established evidence (see Chapter 3). This evidence that high sodium intakes can indirectly mediate risk of adverse health outcomes underpinned the committee's assessment of evidence on associations between sodium intake and direct health outcomes. Taking the evidence for blood pressure effects into consideration as background, the committee focused its review on new evidence on sodium intake and direct health outcomes, particularly evidence from intervention studies where available.

Each outcome is discussed in turn, presenting the available data organized by population group, and within each group organized alphabetically by the last name of the first author. Each study is described by its population, size, and characteristics; study design, purpose, and length; sodium intake measure and method; range of intake, reference intake, and adjustments; outcome measure, confounders, and adjustments; and direction and significance of effect. For each major outcome of CVD, CHF, and CKD, the committee provides a summary table evaluating each study using as criteria the generalizability of the study population to U.S. populations and the appropriateness of the methodology used to support the findings and conclusions.

Finally, a summary of findings and conclusions is given on each major outcome for general populations and for population subgroups of interest as described in the statement of task, specifically those with hypertension or prehypertension, those 51 years of age and older, African Americans, and those with diabetes, CKD, and CHF (see Appendix F for evidence tables).

Studies on General Populations

The committee identified studies related to CVD from its literature search that met the criteria for inclusion described in Appendix F, Table F-1, and, on further examination, were found to be relevant to the committee's task. The cardiovascular health outcomes reviewed were CVD, stroke, stroke mortality or CVD mortality, and all-cause mortality. The committee's summary of the evidence for each cardiovascular health outcome is shown in Tables 4-1 through 4-6.

Cohen et al. (2006)

Population size and characteristics Cohen et al. (2006) obtained data from participants in the National Health and Nutrition Examination Survey (NHANES) II from among the general U.S. population (n=7,154 participants 30-74 years of age). Individuals with self-reported preexisting disease, as well as those who reported being on a low-sodium diet for hypertension, were excluded.

Study design, purpose, and length This secondary analysis of NHANES data was carried out to assess the potential impact of dietary sodium intake on risk of CVD and all-cause mortality over a mean of 13.7 years of follow-up.

Sodium intake measure and method Dietary sodium intake was assessed from a 24-hour dietary recall administered in the NHANES survey. The NHANES II dataset does not include sodium from salt added at the table. In addition, because only 1 day of intake was used, the sodium measure used in this study may not have represented the usual dietary intake of sodium at the individual level.

Range of intake, reference, and adjustments Dietary sodium intake was categorized as <2,300 mg [n=3,443] or ≥2,300 mg [n=3,711]; or by quartiles, corresponding to lowest to highest sodium intakes of <1,645, 1,645-2,359, 2,360-3,345, and ≥3,346 mg per day, using the highest intake quartile as reference; and also as a continuous variable (per 1,000 mg). Sodium intake was adjusted for calories and the highest and lowest 1 percent of the calorie intake range were excluded from the analysis. The correlation between sodium and caloric intake for most age/sex groups is greater than 0.7 (IOM, 2010, pp. 129-130).

Outcome measure, confounders, and adjustments Outcomes measured were CVD mortality, all-cause mortality, coronary heart disease mortality, and cerebrovascular disease mortality. All analyses were adjusted for prior hypertension and systolic blood pressure, which may be in the causal pathway.

Direction and significance of effect For quartiles of sodium intake the lower sodium intake quartiles were not associated with CVD mortality (hazard ratio [HR]=1.31 [confidence interval (CI): 0.90, 1.89] p=0.14), 2 (HR=1.39 [CI: 0.91, 2.11] p=0.11), and 3 (HR=0.89 [CI: 0.64, 1.25] p=0.49). Similarly, no significant associations were found between sodium intake quartiles and all-cause or stroke mortality. However, when analyzed for intakes less than 2,300 mg per day compared to 2,300 mg per day or greater, lower sodium intake was statistically significantly associated with increased risk of all-cause mortality. Models of sodium density, expressed as a sodium-to-calorie ratio, showed a statistically significant inverse association with all-cause mortality (HR=0.89 [CI: 0.79, 1.00] p=0.05). In other words, lower sodium intake was associated with higher all-cause mortality. For dietary sodium intake measured as a continuous variable, a statistically significant inverse relationship was found between sodium intake and CVD mortality whether expressed as sodium per mg (HR=0.89 [CI: 0.80, 0.99] p=0.03) or as sodium per calorie (HR=0.80 [CI: 0.68, 0.94] p=0.008).

Interactions Although data were not shown, the authors reported that they found no evidence of interactions by age, race, or prevalence of diabetes or hypertension.

Cohen et al. (2008)

Population size and characteristics Cohen et al. (2008) analyzed data from participants in the NHANES III survey from the general U.S. population (n=8,699 participants 30 years of age and older). Individuals with self-reported preexisting disease, as well as those who reported being on a low-sodium diet for hypertension, were excluded.

Study design, purpose, and length This secondary analysis of NHANES data was carried out to assess the potential impact of dietary sodium on risk of CVD and all-cause mortality over a mean period of 8.7 years.

Sodium intake measure and method Energy-adjusted dietary sodium intake was estimated from one 24-hour dietary recall. Use of added salt was determined by the responses "does not add," "adds some," or "adds a lot."

Range of intake, reference, and adjustments Sodium intake was categorized into quartiles of <2,060 [n=2,174], 2,060-2,921 [n=2,175], 2,922-4,047 [n=2,175], and 4,048-9,946 mg per day [n=2,175], using the highest intake quartile as reference. Intake analyses were adjusted for added table salt.

Outcome measure, confounders, and adjustments Outcomes included CVD and all-cause mortality and were adjusted for blood pressure and hypertension, which may be in the causal pathway. Other variables examined as potential confounders included sex, age, serum cholesterol, race, treatment for hypertension, blood pressure, smoking, alcohol consumption weight, body mass index (BMI), history of diabetes, education, and added table salt.

Direction and significance of effect When fully adjusted, there was a statistically significant higher risk of CVD mortality (p=0.03) with the lowest vs. the highest quartile of sodium intake that was not present for all-cause mortality (p=0.17). For sodium as a continuous variable and residuals-adjusted sodium, nonsignificant inverse trends were found for both CVD and all-cause mortality. Similar trends were found using sensitivity analyses on the subset of participants (n=5,560) who reported not adding salt during cooking or at the table.

Interactions Although data were not shown, the authors reported that they found no evidence of interactions by age, race, or prevalence of diabetes or hypertension.

Gardener et al. (2012)

Population size and characteristics Gardener et al. (2012) analyzed data from participants in the Northern Manhattan Study (n=2,657) who had no previous diagnosis of stroke, were older than 40 years of age (mean 69 ±10 years) and were ethnically diverse (21 percent white, 24 percent African American, 53 percent Hispanic).

Study design, purpose, and length This multiethnic population-based prospective cohort study examined associations between sodium consumption and risk of stroke and combined vascular events (stroke, myocardial infarction [MI], and vascular death) over a mean of 10 years.

Sodium intake measure and method Dietary sodium intake was estimated using the Block National Cancer Institute food frequency questionnaire (FFQ). Average sodium intake was examined continuously.

Range of intake, reference, and adjustments Sodium intake was calculated from self-reported data and categorized into tertiles of ≤1,500, 1,501-3,999, and ≥4,000 mg per day. The lowest tertile was used as the reference.

Outcome measure, confounders, and adjustments The primary outcome measure was incident stroke of all subtypes; secondary outcomes were confirmed incident of combined vascular event (combined), MI incident, and vascular death. Analytic models were adjusted for demographics, behavioral risk factors, and vascular risk factors (diabetes, hypercholesterolemia, hypertension, and continuous blood pressure measurements).

Direction and significance of effect This study found that sodium intake was positively associated with increased risk of stroke. Using sodium as a continuous variable, stroke risk increased 17 percent for each 500 mg per day higher sodium intake (HR=1.17 [CI: 1.07, 1.27]). However, the authors noted that the relationship did not appear linear. Participants who consumed more than 4,000 mg sodium daily had a 2.5-fold increase in risk of total stroke compared to those who consumed less than 1,500 mg per day (HR=2.50 [CI:1.23, 5.07]). This difference persisted after adjustment for vascular risk factors. Those who consumed more than 1,500 but less than 4,000 mg of sodium daily had an approximate 30 percent increased risk, though this was not statistically significant. Each 500 mg per day higher sodium intake was associated with a 16 percent greater risk of ischemic stroke. Among those who consumed more than 4,000 mg compared to those who consumed less than 1,500 mg per day (reference intake), the risk was 2.4-fold greater. Consumption of more than 4,000 mg per day also was associated with an increased risk of combined vascular events, while the results were less consistent for lower levels of sodium consumption and cardiovascular events.

Interactions Although data were not reported, the authors found no evidence of interactions by age, race, or prevalence of diabetes or hypertension.

Larsson et al. (2008)

Population size and characteristics Larsson et al. (2008) analyzed data on 26,556 male smokers in Finland, 50-69 years of age, who participated in the Alpha-Tocopherol, Beta-Carotene Cancer Prevention primary prevention trial (ATBC Cancer Prevention Study Group, 1994). Participants had smoked at least five cigarettes daily at baseline. From the original cohort, men who had evidence of disease, for example, history of cancer, and those receiving anticoagulant therapy or taking excess vitamin supplements were excluded from the analysis, although those with a history of diabetes or coronary heart disease were included.

Study design, purpose, and length This large prospective study followed a male cohort to assess relationships between magnesium, calcium, potassium, and sodium intake and risk of stroke for a mean of 13.6 years.

Sodium intake measure and method Energy-adjusted dietary sodium intake was estimated from a self-administered, validated 276-item FFQ that included items commonly consumed in Finland. Use of cooking salt was included in the questionnaire, whereas salt added at the table was not captured.

Range of intake, reference, and adjustments Median sodium intake estimates were divided into quintiles with means of 3,909, 4,438, 4,810, 5,212, and 5,848 mg per day, for Q1 through Q5, respectively, using the lowest quintile as the reference. Thus, the average sodium intake in the United States would be within the lowest quintile of this study.

Outcome measure, confounders, and adjustments Outcome measure was first-ever stroke occurring between the date of randomization (between 1985 and 1988) and the first occurrence of stroke by the end of the study period (December 31, 2004). Strokes were classified as cerebral infarction, intracerebral hemorrhage, subarachnoid hemorrhage, or unspecified stroke, and identified from the Finnish National Hospital Discharge Register and the National Register of Causes of Death. Multivariate analysis adjusted for age, supplementation group, number of cigarettes smoked daily, BMI, systolic and diastolic blood pressure, serum total cholesterol and high-density cholesterol, as well as history of diabetes and coronary heart disease, leisure-time physical activity, and intake of alcohol and total energy.

Direction and significance of effect The analyses found no significant association between dietary sodium intake and risk of any stroke subtype (p for trend for multivariate relative risk [RR]=0.99, 0.06, and 0.55 for cerebral infarction, intracerebral hemorrhage, and subarachnoid hemorrhage, respectively).

Nagata et al. (2004)

Population size and characteristics Nagata et al. (2004) analyzed a cohort of Japanese adults 35 years of age and older using data collected in the Takayama population-based study. The study included 13,355 men and 15,724 women.

Study design, purpose, and length This population-based cohort study examined associations between dietary sodium intake and stroke mortality risk over a period of 7 years.

Sodium intake measure and method Dietary sodium intake was estimated from a 169-item semiquantitative FFQ. The instrument included questions on use of table salt and salty condiments. Use of cooking salt was not included in the questionnaire.

Range of intake, reference, and adjustments Sodium intake levels were represented by tertile for men (means 4,070, 5,209, and 6,613 mg per day) and women (means 3,799, 4,801, and 5,930 mg per day), using the lowest tertile as the reference. Sodium intake was energy-adjusted when used in analyses. Thus, the average sodium intake in the United States would be within the lowest tertile of this study.

Outcome measure, confounders, and adjustments The outcome measure was stroke mortality (data obtained from National Vital Statistics for Japan). Stroke types included subarachnoid hemorrhage, intracerebral hemorrhage, ischemic stroke, and stroke of undetermined type. Adjustment variables included age, education level, marital status, BMI, smoking, alcohol use, and history of diabetes or hypertension.

Direction and significance of effect Over the 7-year follow-up period, 43 subarachnoid hemorrhages, 59 intracerebral hemorrhage deaths, and 137 ischemic stroke deaths occurred. Among men, a 2.3-fold increased risk of stroke mortality (significantly positive for intracerebral hemorrhage and ischemic stroke death) was associated with the highest tertile of sodium intake (mean intake of 7,194 mg per day) after adjustment for other dietary variables (HR=2.33 [CI: 1.23, 4.45] p for trend=0.009). The authors also reported a borderline significant trend between high sodium intake (mean intake of 6,478 mg per day) and total stroke and ischemic stroke death among women (total stroke HR=1.70 [CI: 0.96, 3.02] p for trend=0.07); ischemic stroke (HR=2.10 [CI: 0.96, 4.62] p for trend=0.05).

Stolarz-Skrzypek et al. (2011)

Population size and characteristics Stolarz-Skrzypek et al. (2011) obtained data from two population-based prospective cohort studies (the Flemish Study on Environment, Genes, and Health Outcomes and the European Project on Genes in Hypertension) to examine a cohort of 2,856 participants recruited from a random sample of households in several European countries. The cohort included men and women 20-39, 40-59, and 60 or more years of age.

Study design, purpose, and length This study prospectively examined associations between sodium and changes in blood pressure and risk of CVD mortality and all-cause mortality over a median of 7.9 years.

Sodium intake measure and method Timed 24-hour urine samples were collected 1 week following blood pressure measurements and analyzed for sodium and potassium.

Range of intake, reference, and adjustments Twenty-four-hour sodium excretion values were categorized into tertiles of low (50-126 mmol [1,150-2,898 mg] for women; 50-158 mmol [1,150-3,634 mg] for men); medium (127-177 mmol [2,921-4,071 mg] for women; 159-221 mmol [3,657-5,083 mg] for men); and high (178-400 mmol [4,094-9,200 mg] for women; 222-400 mmol [5,106-9,200 mg] for men).

Outcome measure, confounders, and adjustments Risk of CVD events was assessed for each quartile against risk of CVD events for the whole study population. Cardiovascular outcomes, fatal and nonfatal stroke, fatal and nonfatal MI, fatal and nonfatal left ventricular heart failure, aortic aneurysm, cor pulmonale, and pulmonary or arterial embolism were validated by physicians against medical records. Covariables in the regression analyses were study population, sex, age, blood pressure level, BMI, alcohol intake, use of antihypertensive drugs, urinary potassium excretion, education, smoking status, total cholesterol, and diabetes.

Direction and significance of effect Overall, the authors found that lower sodium intake was associated with higher risk of CVD mortality. In the low, medium, and high tertiles of sodium excretion the CVD mortalities were 50 (4.1% [CI: 3.5%, 4.7%]); 24 (1.9% [CI: 1.5%, 2.3%]); and 10 (0.8% [CI: 0.5%, 1.1%]) events, respectively, after adjustment for risk factors, including baseline hypertension and blood pressure level. The risk of CVD mortality was statistically significantly higher in the low versus the high tertile (HR=1.56 [CI: 1.02, 2.36] p=0.04) with a significant trend over tertiles (p=0.02). All-cause mortality showed a trend in risk of CVD mortality for low and medium tertiles, although it was not statistically significant (HR=1.14 [CI: 0.87, 1.50] and 64 (HR=0.94 [CI: 0.75-1.18]). Likewise, there was no significant effect on total CVD incidence. Analysis of supplemental data from this study demonstrated that individuals assigned to the low-sodium tertiles were older, less educated, had greater comorbidities, lower urine volume, and lower serum creatinine than those in higher-sodium tertiles.

Takachi et al. (2010)

Population size and characteristics Takachi et al. (2010) examined data from the Japan Public Health Center-based Prospective Study, conducted in two cohorts. Cohort I and II participants were 40-59 and 40-69 years of age, respectively. Those with a history of cancer or coronary heart disease were excluded, leaving a final study population of 77,500 (35,730 men and 41,770 women).

Study design, purpose, and length The objective of this prospective study was to assess associations between sodium and salted food consumption and risk of either cancer or CVD. Participants were followed from the beginning in 1990 (cohort I) or 1993 (cohort II) until December 31, 2004.

Sodium intake measure and method Dietary sodium intake data were determined from a 138-item FFQ that included cooking salt, soy sauce, table salt, and other salty condiments.

Range of intake, reference, and adjustments Energy-adjusted sodium intake per day was categorized by quintile: medians were 3,084, 4,005, 4,709, 5,503, and 6,844 mg per day for Q1 through Q5, respectively. Thus, the average sodium intake in the United States would be close to the lowest quintile of this study.

Outcome measure, confounders, and adjustments Cardiovascular outcomes included diagnosis of MI and diagnosis of stroke confirmed by computed tomography (CT) scan and/or magnetic resonance imaging (MRI) from medical records. Adjustment variables for analysis were sex and age, with additional adjustment for BMI, smoking status, alcohol consumption, physical activity, and quintiles of energy, potassium, and calcium.

Direction and significance of effect Adjusted multivariate analysis found a significant positive association between sodium consumption at the highest compared to the lowest quintile and risk of stroke (HR=1.21 [CI: 1.01, 1.43] p for trend=0.03) and between use of cooking and table salt and risk of stroke (HR=1.21 [CI: 1.02, 1.44] p for trend=0.05), but not between use of dried salted fish and risk of stroke. Increased intake of dried and salted fish was associated with lower risk of MI. The risk of the composite CVD endpoint was elevated in the highest quintile of sodium (HR=1.19 [CI: 1.01, 1.40] p for trend=0.06). The results also showed correlation with other variables, such as dried and salted fish, although the impact of those variables on the outcomes is unknown.

Umesawa et al. (2008)

Population size and characteristics Umesawa et al. (2008) examined data from participants in the Japan Collaborative Cohort Study for Evaluation of Cancer Risks (23,119 men and 35,611 women) 40-79 years of age.

Study design, purpose, and length This large prospective study examined associations between dietary sodium intake and mortality from stroke; stroke related to subarachnoid hemorrhage, intraparenchymal hemorrhage, or ischemic stroke; coronary heart disease; and total CVD over a mean follow-up period of 12.7 years.

Sodium intake measure and method Energy-adjusted dietary sodium intake was estimated from responses to a 35-item FFQ and the results were calibrated against a previous validation study that included items from the FFQ and four 3-day dietary records.

Range of intake, reference, and adjustments Dietary sodium intake was categorized into quintiles from lowest to highest with Q1=101 ±30 (2,323 ±690), Q2=146 ±11 (3,358 ±253), Q3=182 ±11 (4,186 ±253), Q4=220 ±12 (5,060 ±276), and Q5=272 ±36 (6,256 ±828) mmol (mg) per day. The calibrated sodium intake based on a validation study was twice as high in all five quintiles in which median values were assigned for each quintile and the significance of the variables was tested.

Outcome measure, confounders, and adjustments Outcome measures were stroke (including ischemic stroke), CVD, or CHD mortality derived from data obtained from death certificates for targeted populations in each study locale. Adjustments were made for CVD risk factors, including hypertension, and for potassium intake.

Direction and significance of effect The authors found an association between greater dietary sodium intake and greater mortality from total stroke, ischemic stroke, and total CVD. Multivariable hazard ratios were strongest with the highest compared to the lowest quintiles of sodium intake: HR=1.55 (CI: 1.21, 2.00) for total stroke, HR=2.04 (CI: 1.41, 2.94) for ischemic stroke, and HR=1.42 (CI: 1.20, 1.69) for total CVD mortality. The positive associations found between dietary sodium intake and mortality risk of total and ischemic stroke and CVD were independent of potassium intake and body weight. No significant association was found between sodium intake and risk of CHD.

Yang et al. (2011)

Population size and characteristics Yang et al. (2011) analyzed data from 12,267 adults 20 years of age and older from the NHANES III. NHANES participants were linked to mortality data from the National Death Index through December 2006.

Study design, purpose, and length This secondary analysis of NHANES III data was carried out to examine associations between sodium intake, potassium intake, the sodium-to-potassium ratio, and risk of CVD mortality and all-cause mortality over an average of 14.8 years.

Sodium intake measure and method Estimates of dietary sodium intake were derived from the NHANES 24-hour dietary recall. Within-person variability was calculated using 7 percent of individuals and this variability was used to adjust the sodium intake.

TABLE 4-1 Weaknesses and Strengths of Population Studies and Methods of Studies on Cardiovascular Health Outcomes: CVD Outcomes in General Populations

Study	Sample Size and Population	Study Design	Method to Assess Sodium Intake
Cohen et al., 2006	*Strengths*	*Strengths*	*Weaknesses*
	Good generalizability to the general population (U.S.)	Prospective cohort	24-h recall, only one day recorded

Range of intake, reference, and adjustments Sodium intake levels, categorized into quartiles from lowest to highest are Q1=2,176; Q2=3,040; Q3=3,864; and Q4=5,135 mg per day, using the lowest intake as reference. Nutrient disease associations were estimated as continuous variables for all-cause and CVD mortality. Because relationships between estimated usual intakes and all-cause and CVD mortality were approximately linear, the percentile distributions of estimated usual intakes were calculated as the middle value of each quartile. Usual intake estimates were derived using a method adjusting for within-person variation developed by the National Cancer Institute (Tooze et al., 2006). Usual sodium (and potassium) intake estimates for CVD mortality, all-cause mortality, and ischemic heart disease (IHD) mortality were reported.

Outcome measure, confounders, and adjustments Outcome measures were CVD (including IHD) mortality and all-cause mortality. Unlike Cohen et al. (2008), the investigators elected not to adjust for measured blood pressure or antihypertensive treatment, arguing that these factors may be in the causal pathway linking dietary sodium to health outcomes.

Sodium Intake Levels or Intake Ranges	Adjustment for Confounders	Other
Strengths	*Strengths*	*Strengths*
<1,645-≥3,346 mg/d	Adjustments for caloric intake, education, BMI, tobacco and alcohol use	Eliminated patients with self-reported CVD and those on low-Na diet for medical reasons (reverse causation)
	Na intake adjusted for DM	Eliminated highest and lowest 1% of Na and calories (measurement error)
	Weaknesses	
	Adjusted for blood pressure and hypertension, which could be in the causal pathway	*Weaknesses*
		Inconsistency of BMI and reported calories

continued

TABLE 4-1 Continued

Study	Sample Size and Population	Study Design	Method to Assess Sodium Intake
Cohen et al., 2008	*Strengths*	*Strengths*	*Weaknesses*
	Good generalizability to the general population (U.S.)	Prospective cohort	24-h recall, only 1 day recorded Na intake adjusted for added salt instead of including in exposure
Gardener et al., 2012	*Strengths*	*Strengths*	*Strengths*
	Good generalizability to the general population (U.S.)	Population-based prospective cohort	Na consumption assessed over previous year; questionnaire modified for Hispanic food items *Weaknesses* FFQ may not be well calibrated; FFQ was not able to fully capture the contribution of salt added to foods at the table

Sodium Intake Levels or Intake Ranges	Adjustment for Confounders	Other
Strengths	*Strengths*	*Strengths*
2,060-4,048 mg/d	Adjustment for caloric intake	Eliminated patients with self-reported CVD and those on low-Na diet for medical reasons (reverse causation)
	Na intake adjusted for DM, cancer, education, and tobacco and alcohol use	Eliminated highest and lowest 1% of Na and caloric intake (measurement error)
	Weaknesses	*Weaknesses*
	Adjusted for blood pressure and hypertension, which could be in the causal pathway	Caloric intake could be underreported in individuals with low Na intake, leading to a systematic error
Strengths	*Strengths*	*Strengths*
≤1,500-10,000 mg/d	Analysis conducted with and without adjustment for vascular risk factors	Eliminated patients with extreme Na and caloric intake and those with stroke and MI

continued

TABLE 4-1 Continued

Study	Sample Size and Population	Study Design	Method to Assess Sodium Intake
Larsson et al., 2008	*Weaknesses*	*Strengths*	*Strengths*
	Limited generalizability to the general population (male smokers, Finland)	Prospective cohort	276 items in FFQ
			Cooking salt included
			Weaknesses
			FFQ may not be well calibrated
			Table salt not included
Nagata et al., 2004	*Strengths*	*Strengths*	*Strengths*
	Good generalizability to the general population (Japan)	Prospective cohort	169 items in FFQ
			Added salt considered
Stolarz-Skrzypek et al., 2011	*Strengths*	*Strengths*	*Strengths*
	Good generalizability to the general population (Dutch)	Population-base prospective cohort	UNa collection
			Elimination of low 24-h urine volume and extreme 24-h urine creatinine 24-h urine collection
			Weaknesses
			Creatinine data suggest undercollection of urine specimens in lowest tertile
			No assessment of dietary intake and thus, no ability to adjust for caloric intake

Sodium Intake Levels or Intake Ranges	Adjustment for Confounders	Other
Strengths	*Strengths*	*Strengths*
3,909-5,848 mg/d	Analysis conducted with and without adjustments for BP	Eliminated patients with self-reported CVD
	Na intake adjusted for caloric intake, BMI, DM CHD, and tobacco and alcohol use	
Weaknesses	*Strengths*	*Strengths*
High Na intake of questionable relevance to the U.S. population (4,070-6,613 and 3,799-5,930 mg/d for men and women, respectively)	Na intake adjusted for caloric intake using residual method, BMI, and tobacco and alcohol use	Eliminated patients with self-reported CVD and cancer (reverse causation)
	Na intake adjusted for education	
Strengths	*Strengths*	*Weaknesses*
2,461-5,980 mg/d	Na intake adjusted for urinary K, BMI, DM, education, and tobacco and alcohol use	Individuals in lower Na tertiles were older and less educated
	Weaknesses	
	Adjusted for blood pressure and hypertension, which may be in the causal pathway	

continued

TABLE 4-1 Continued

Study	Sample Size and Population	Study Design	Method to Assess Sodium Intake
Takachi et al., 2010	*Strengths*	*Strengths*	*Strengths*
	Good generalizability to the general population (Japan)	Prospective cohort	138 items in FFQ
			Cooking salt and table salt included
			Validated FFQ estimates with 28-d diet records and 24-h urines with correlations of 0.30-0.50
			Reproducibility examined by repeat FFQs with correlation of 0.49-0.67
Umesawa et al., 2008	*Strengths*	*Strengths*	*Weaknesses*
	Good generalizability to the general population (Japan)	Prospective cohort	Only 35 items in FFQ (question of validity)
			Estimated mean Na intake was half that of diet records and needed recalibration
Yang et al., 2011	*Strengths*	*Strengths*	*Weaknesses*
	Good generalizability to the general population (U.S.)	Prospective cohort	24-h recall, only one day recorded
			Using 7.4% of the participants with second day 24-h recalls to estimate usual sodium intake accounting for within-person variation in intake

NOTES: Sodium intake presented as mmol in a study was converted to mg using 23 mg/mmol. BMI, body mass index; BP, blood pressure; CHD, coronary heart disease; CVD, cardiovascular disease; d, day; DM, diabetes mellitus; FFQ, food frequency questionnaire; h, hour; K, potassium; mg, milligram; MI, myocardial infarction; Na, sodium, UNa, urinary sodium.

Sodium Intake Levels or Intake Ranges	Adjustment for Confounders	Other
Strengths	*Strengths*	*Strengths*
3,084-6,844 mg/d	Na intake adjusted for caloric intake, BMI, and tobacco and alcohol use	Eliminated patients with self-reported CVD and cancer (reverse causation) and those with extreme caloric intake
Strengths	*Strengths*	*Strengths*
2,323-6,256 mg/d	Adjustment for K intake Na intake adjusted for DM, caloric intake, BMI, tobacco and alcohol use, and education	Eliminated patients with self-reported CVD and cancer
Strengths	*Strengths*	
2,176-5,135 mg/d for the 12.5 and 87.5 percentiles Intake range=839-8,555 mg/d	Conducted analysis with and without adjustment for BP Na intake adjusted for family history of CVD, education, caloric intake, and tobacco and alcohol use Na intake adjusted for education Na intake adjusted for DM	

Direction and significance of effect After multivariable adjustment, higher usual sodium intake was found to be directly associated with all-cause mortality (HR=1.20 [CI: 1.03, 1.41] per 1,000 mg per day), but not with CVD mortality or IHD mortality. However, the finding that correction for regression dilution increased the effect on all-cause mortality, but not on CVD mortality, is inconsistent with the theoretical causal pathway.

Interactions Although data were not reported, the authors found no evidence of interactions by age, race, or prevalence of diabetes or hypertension.

Studies in Populations 51 Years of Age and Older

Geleijnse et al. (2007)

Population size and characteristics Geleijnse et al. (2007) examined a subset of data from the older cohort of the Rotterdam prospective cohort study that included 1,448 randomly selected participants 55 years of age and older living in the Netherlands.

Study design, purpose, and length This case-cohort design examined the relationships between sodium and potassium intake and incidence of MI

TABLE 4-2 Weaknesses and Strengths of Population Study and Methods of Studies on Cardiovascular Health Outcomes: CVD Outcomes in Populations 51 Years of Age and Older*

Study	Sample Size and Population	Study Design	Method to Assess Sodium Intake
Geleijnse et al., 2007	*Strengths*	*Strengths*	*Weaknesses*
	Good generalizability to the general population (Dutch)	Population-based prospective case-cohort study	Single overnight urine collection (question of validity)

* Five studies analyzed the data on health outcomes by age and found no interactions (Cohen et al., 2006, 2008; Cook et al., 2007; Gardener et al., 2012; Yang et al., 2011). NOTES: Sodium intake presented as mmol in a study was converted to mg using 23 mg/ mmol. BMI, body mass index; CVD, cardiovascular disease; DM, diabetes mellitus; h, hour; K, potassium; SD, standard deviation; UNa, urinary sodium.

and stroke and of CVD mortality and all-cause mortality. Participants were followed for a median of 5.5 years.

Sodium intake measure and method Urinary sodium, potassium, and creatinine excretion were estimated from a single overnight urine sample collected at home.

Range of intake, reference, and adjustments In a sub-cohort free of CVD and hypertension, 24-hour urinary sodium was divided into quartiles for analysis of all-cause mortality. Quartile levels were 66, 105, and 151 mmol (1,518, 2,415, and 3,473 mg) per day. Adjustments were made for total energy, alcohol intake, saturated fat intake, and 24-hour urinary potassium.

Outcome measure, confounders, and adjustments Outcome measures were CVD mortality and all-cause mortality. Mean baseline blood pressure for the cohort was 140 (standard deviation [SD]=22) mmHg systolic and 74 (SD=11) mmHg diastolic; 37 percent had a diagnosis of hypertension. Adjustments were made for age, sex, 24-hour urinary creatinine, BMI, smoking status, diabetes, use of diuretics, and education.

Sodium Intake Levels or Intake Ranges	Adjustment for Confounders	Other
1 SD increases in UNa excretion	*Strengths*	*Strengths*
	Adjusted for 24-h urinary creatinine excretion; BMI; smoking status; DM; use of diuretics; highest completed education; dietary confounders (intake of total energy, alcohol, calcium, sat. fat); K excretion	Elimination of patients with CVD and hypertension

Direction and significance of effect This study found no significant difference between urinary sodium level and risk of CVD mortality or all-cause mortality. Using the lower quartile as the reference, the study found an inverse association with CVD mortality that was borderline statistically significant (RR=0.77 [CI: 0.60, 1.01] per 1 SD). After excluding participants with a history of CVD or hypertension, the difference was attenuated and nonsignificant. In an examination across quartiles of urinary sodium level, using the lowest quartile as reference, all-cause mortality in disease-free participants relative risk by quartile were 0.80 (CI: 0.43,1.49), 0.66 (CI: 0.34, 1.27), and 0.98 (CI: 0.54, 1.78), respectively.

Studies with Additional Analysis of Data by Age

Five of the nine reported studies in the general population listed above also analyzed the data on health outcomes by age and found no interaction (Cohen et al., 2006, 2008; Cook et al., 2007; Gardener et al., 2012; Yang et al., 2011).

Studies on Populations with Chronic Kidney Disease

Dong et al. (2010)

Population size and characteristics Dong et al. (2010) collected data on 305 incident peritoneal dialysis patients (with diabetes or with preexisting CVD) in Japan. The cohort included 129 men and 176 women with a mean age of 59.4 years.

Study design, purpose, and length The purpose of this retrospective cohort study was to determine whether dietary sodium intake was correlated with CVD mortality and all-cause mortality over a period of 31.4 (±13.7) months.

Sodium intake measure and method Dietary sodium intake was estimated by a dietitian from 3-day diet records that also asked about use of added salt.

Range of intake, reference, and adjustments Sodium intake was categorized by tertile: mean 1,410 (±0.17), 1,810 (±0.11), and 2,470 (±0.54) mg per day for low, middle, and high tertiles, respectively (sodium removal, dialysate + urine=2,200 [±1,210], 2,780 [±1,090], and 3,030 [±1,100] mg per day for low, middle, and high tertiles, respectively). Statistical comparisons were made among intake tertiles. Sodium intake during the first 3 months was represented as the baseline sodium intake. Daily total protein and daily energy intakes were normalized for standard body weight. All of the measurements during the study were averaged.

Outcome measure, confounders, and adjustments Outcome measures were all-cause and CVD mortality. Covariates for baseline sodium intake included age, sex, BMI, history of diabetes, CVD, and biochemical measures.

Direction and significance of effect The authors found that the lowest sodium intake was associated with increased mortality risk. Mean baseline systolic blood pressure was similar across tertiles of sodium intake. Compared to the highest sodium intake tertile, those in the lowest tertile were at 55 percent lower risk of all-cause mortality (p=0.02), and 67 percent lower risk of CVD mortality (p=0.07). Similarly, when analyzed as a continuous variable, average sodium intake was correlated with overall mortality (HR=0.44 [CI: 0.20, 0.95] p=0.04), and CVD mortality (HR=0.11 [CI: 0.03, 0.48] p=0.003), suggesting that average low sodium intake was a predictor of CVD mortality.

Heerspink et al. (2012)

Population size and characteristics Heerspink et al. (2012) examined participants in the Reduction of Endpoints in NIDDM [non-insulin-dependent diabetes mellitus] with the Angiotensin II Antagonist Losartan study (250 centers in 28 countries in Asia, Europe, and the Americas) and the Irbesartan Diabetic Nephropathy Trial (209 centers in the Americas, Australia, Europe, and Israel). Participants were 30-70 years of age, with type 2 diabetes and overt proteinuria.

Study design, purpose, and length This prospective cohort study evaluated whether the effect of randomization to treatment with angiotensin receptor blockers (ARBs) with kidney disease progression and CVD was modified by dietary sodium intake over a period of 30 months.

Sodium intake measure and method Dietary sodium intake was assessed using multiple (average of five collections per participant) 24-hour urinary excretion samples collected throughout the study. The 24-hour urine sodium measure was indexed to creatinine in an attempt to control for quality of collections.

Range of intake, reference, and adjustments The 24-hour sodium-to-creatinine ratio tertiles were <121, 121-153, and >153 mmol/g (<2,783, 2,783-3,519, and >3,519 mg/g, respectively). Sensitivity analyses were conducted using 24-hour urine sodium without indexing to urine creatinine.

Outcome measure, confounders, and adjustments The outcomes were CKD progression, defined as doubling of serum creatinine or incident end-stage renal disease (ESRD), and CVD, defined as a composite of CVD death, MI, stroke, hospitalization for CHF, or revascularization procedure.

Direction and significance of effect Results from this study suggest that ARBs were more effective at decreasing CKD progression and CVD when sodium intake was in the lowest tertile [<121 mmol per day (<2,783 mg per day)].

TABLE 4-3 Weaknesses and Strengths of Population Studies and Methods of Studies on Cardiovascular Health Outcomes: CVD Outcomes in CKD Populations

Study	Sample Size and Population	Study Design	Method to Assess Sodium Intake
Dong et al., 2010	*Weaknesses*	*Weaknesses*	*Strengths*
	Poor generalizability to subgroups of interest (men and women receiving peritoneal dialysis, China)	Retrospective cohort	Repeated 3-d diet records taken over 3 mo
Heerspink et al., 2012	*Weaknesses*	*Strengths*	*Strengths*
	Poor generalizability to subgroups of interest (men and women with type 2 diabetes nephropathy)	Prospective cohort study	Used average of multiple 24-h urine collection measurements 24-h urine collection corrected by creatinine

NOTES: Sodium intake presented as mmol in a study was converted to mg using 23 mg/mmol. BMI, body mass index; CVD, cardiovascular disease; d, day; DM, diabetes mellitus; h, hour; LDL, low-density lipoprotein cholesterol; mg, milligram; mo, month; Na, sodium; Kt/V, measurement of urea removal.

Studies on Populations with Cardiovascular Disease

Costa et al. (2012)

Population size and characteristics Costa et al. (2012) followed a subset of 372 participants in the Brasilia Heart Study. This study included only individuals who had experienced a recent (within 24 hours) MI and who sought medical care, not individuals from the general population.

Study design, purpose, and length This prospective cohort study was carried out to examine the influence of high vs. low sodium intake on inflammatory-oxidative response, cardiac remodeling, and total mortality after MI, for up to 4 years. Cardiac MRI was used to identify cardiac

Sodium Intake Levels or Intake Ranges	Adjustment for Confounders	Other
Strengths	*Strengths*	*Weaknesses*
760-5,530 mg/d	Adjustments for BMI, history of DM or CVD, baseline total Kt/V, total creatinine clearance, mean arterial pressure, serum albumin, hemoglobin, calcium × phosphate, LDL, C-reactive protein	Selected populations on a medically advised low-Na diet at inclusion
Weaknesses		
Na and energy intake markedly lower than the average levels for the general population in Beijing		Observations in those with low Na intake suggest reverse causation (Na intake may have been affected by acute illness)
		Small sample
Strengths	*Weaknesses*	*Weaknesses*
<2,783-≥3,519 mg/d	None	Study not designed to look at urine Na but at the interaction with angiotensin receptor blockers

changes and areas of infarction. Study participants received lifestyle counseling upon discharge, which included diet counseling.

Sodium intake measure and method Dietary sodium intake was estimated
from a 62-item validated FFQ.

Range of intake, reference, and adjustments Energy-adjusted sodium intake
was categorized as either high ($\geq$1,200 mg per day) or low (<1,200 mg
per day). A validation study of 24-hour urine excretions in 21 patients
indicated that values were consistently 1,500 mg ±500 mg greater than
the estimated daily sodium intake from the questionnaire. Sodium intake
was assessed for associations with alcohol intake, smoking, income, and
education level.

Outcome measure, confounders, and adjustments The primary outcome
was total mortality in the first 30 days following MI. Secondary outcomes
were total mortality during 4-year follow-up, composite endpoint of fatal
or nonfatal MI, and unstable angina. Outcomes were recorded from 48
hours following onset of symptoms of MI. Associations with outcomes
were tested for potential confounders, namely blood pressure changes during hospitalization, use of antihypertensive medication, and presence of
pulmonary congestion.

Direction and significance of effect Significant correlations were found
between sodium intake and percentage of fat and calories in daily intake. In
the first 30 days following MI, total mortality was statistically significantly
higher in the high- compared to the low-sodium groups (p=0.04), and the
association remained similar after excluding non-CVD-related mortality
(p=0.02). Multivariate analyses found a 2.9-fold risk of mortality with
high sodium intake. Overall, for the first 30 days and up to 4 years after-
ward, total mortality was significantly associated with high sodium intake
(p<0.05).

Kono et al. (2011)

Population size and characteristics Kono et al. (2011) examined a Japanese
cohort of 78 men and 24 women, with a mean age of 64 years. All participants had a recent (within 2 weeks) hospitalization for acute ischemic
stroke and thus the study did not evaluate participants from the general
population. Both initial and later vascular events were confirmed by a neu-
rologist using clinical data, including CT scan or MRI. Participants were
divided into groups with large vessel disease (LVD) or small vessel disease
(SVD).

Study design, purpose, and length This prospective cohort study examined the association of timed urine sodium excretion with recurrence rates and risk of stroke and other vascular events, including MI, angina pectoris, and peripheral artery disease over a period of 3 years.

Sodium intake measure and method For 3 consecutive days, participants collected urine for 8 hours overnight using a self-monitoring device.

Range of intake, reference, and adjustments Mean daily sodium intake was calculated for each participant, divided into four groups, and analyzed according to median value of physical activity and sodium intake. For analysis, sodium intake was expressed as greater or less than 3,201 mg per 24 hours.

Outcome measure, confounders, and adjustments Outcome measures were cumulative recurrence rates of stroke. Adjustments were made for age and medications. Deaths without stroke recurrence were excluded from the analysis.

Direction and significance of effect Cumulative risk analysis found that a salt intake of greater than the median of 10,700 mg per day (4,280 mg of sodium) was associated with higher stroke recurrence rate (HR=2.43 [CI: 1.04, 5.68] p=0.04). Univariate analysis of lifestyle management also found that poor lifestyle, defined by both high salt intake ($\geq$10,700 mg [4,000 mg sodium] per day) and low physical activity (<5,800 steps per day), was significantly associated with stroke recurrence (HR=1.71 [CI: 1.11, 2.62] p=0.013).

O'Donnell et al. (2011)

Population size and characteristics O'Donnell et al. (2011) reanalyzed data from two randomized controlled drug trials, the ONgoing Telmis-artan Alone and in combination with Ramipril Global Endpoint Trial (ONTARGET) and the Telmisartan Randomized AssessmeNt Study in aCe iNtolerant subjects with cardiovascular Disease (TRANSCEND). Both tri-als included participants with established CVD. The combined study sample had a total of 28,880 participants, 55 years of age and older, recruited from 40 countries. A subset of the study cohort (n=2,625) provided a 2-year follow-up and final urinary measurement.

Study design, purpose, and length This study was a prospective observational analysis of two cohorts followed for a median of 56 months (4.7 years).

Sodium intake measure and method Sodium intake was estimated from spot urine sodium measurements from a single morning fasting urine sample collected before the run-in period of the trials. Estimates were extrapolated to 24-hour estimated sodium excretion using the Kawasaki formula (see Chapter 2 for further discussion). This formula was developed in a healthy Asian population with higher dietary sodium intake than the usual intake in the United States. However, O'Donnell et al. (2011) validated the sodium measure in 105 individuals in the Prospective Urban Rural Epidemiology study, and reported correlations of estimated and measured 24-hour urine sodium of 0.55.

Range of intake, reference, and adjustments Sodium excretion was categorized into seven ranges: <2,000, 2,000-2,990, 3,000-3,990, 4,000-5,990, 6,000-6,990, 7,000-8,000, and >8,000 mg per day. The reference sodium excretion was 4,000-5,990 mg per day.

Outcome measure, confounders, and adjustments Risk estimates were made for each outcome (CVD-related mortality, MI, stroke, and hospitalization for CHF) and for a composite of all outcomes. Models for analysis were adjusted for age, sex, race/ethnicity, prior history of stroke or MI, creati-

TABLE 4-4 Weaknesses and Strengths of Population Studies and Methods of Studies on Cardiovascular Health Outcomes: CVD Outcomes in Populations with Preexisting CVD

Study	Sample Size and Population	Study Design	Method to Assess Sodium Intake
Costa et al., 2012	*Weaknesses*	*Strengths*	*Weaknesses*
	Poor generalizability of subgroups of interest (men and women with acute MI, Brazil)	Prospective cohort	Only 62 items in FFQ
			Questionnaire validation conducted among volunteers without MI and may be misleading
			FFQ assessing last 90 d administered within 24 h after MI

nine, BMI, comorbid vascular risk factors (hypertension, diabetes, atrial fibrillation, smoking, low-density lipoprotein, high-density lipoprotein, drug treatment, fruit and vegetable consumption, physical activity, baseline blood pressure, change in systolic blood pressure, and urinary potassium).

Direction and significance of effect For the composite outcome, multivariate analysis found a U-shaped relationship between 24-hour urine sodium and the composite outcome of CVD death, MI, stroke, and hospitalization for CHF. Compared with the reference sodium excretion (4,000-5,990 mg per day), excretion of 7,000-8,000 (HR=1.15 [CI: 1.00, 1.32]) and >8,000 mg per day (HR=1.49 [CI: 1.28, 1.75]), and excretion of 2,000-2,990 (HR=1.16 [CI: 1.04, 1.28]) and <2,000 mg per day (HR=1.21 [CI: 1.04, 1.43]) were associated with increased risk of composite CVD and mortality. Increased risk of CVD-related mortality was associated with excretion of 7,000-8,000 mg per day (HR=1.53 [CI: 1.26, 1.86]), excretion of >8,000 mg per day (HR=1.66 [CI: 1.31, 2.10]), excretion of 2,000-2,990 mg per day (HR=1.19 [CI: 1.02, 1.39]), and excretion of <2,000 mg per day (HR=1.37 [CI: 1.09, 1.73]) compared to the reference excretion. Compared with the reference excretion, excretion of 6,000-6,990 mg per day (HR=1.21 [CI: 1.03, 1.43]) was associated with increased risk of MI, while excretion of >8,000 mg per day was associated with increased risk of stroke (HR=1.48 [CI: 1.09, 2.01]) and hospitalization for CHF (HR=1.51 [CI: 1.12, 2.05]). Excretion of 2,000-2,990 mg per day was associated with increased risk of hospitalization for CHF (HR=1.23 [CI: 1.01, 1.49]). Sub-

Sodium Intake Levels or Intake Ranges	Adjustment for Confounders	Other
Strengths	*Weaknesses*	*Strengths*
≥1,200 or <1,200 mg/d	Only a few potential confounders included (age, sex, hypertension, diabetes, sedentary activity level, BMI)	Additional analysis adjusted Na intake for caloric intake with similar results

continued

TABLE 4-4 Continued

Study	Sample Size and Population	Study Design	Method to Assess Sodium Intake
Kono et al., 2011	*Weaknesses*	*Strengths*	*Strengths*
	Poor generalizability to subgroups of interest (men and women with acute IS, Japan)	Hospital-based cohort	UNa collection; overnight urines averaged over 3 d Weaknesses Sample with acute IS only single overnight urine collection (question of validity)
O'Donnell et al., 2011	*Strengths*	*Strengths*	*Weaknesses*
	Large sample size Good generalizability to subgroups of interest (population at high risk of CVD or DM, Canada)	Follow-up of 2 RCTs of anti-hypertensive agents	Only single overnight urine collection Used Kawasaki formula from general Asian population to estimate 24-h excretion, likely miscalibrated

NOTES: Sodium intake presented as mmol in a study was converted to mg using 23 mg/mmol. BMI, body mass index; BP, blood pressure; CVD, cardiovascular disease; d, day; DM, diabetes mellitus; FFQ, food frequency questionnaire; h, hour; HDL, high-density lipoprotein cholesterol; IS, ischemic stroke; LDL, low-density lipoprotein cholesterol; mg, milligrams; MI, myocardial infarction; Na, sodium; RCT, randomized controlled trial; UK, urinary potassium; UNa, urinary sodium.

Sodium Intake Levels or Intake Ranges	Adjustment for Confounders	Other
Strengths	*Strengths*	*Strengths*
Salt >10,700 mg/d or ≤10,700 mg salt (4,280 mg/Na)/d	Adjustments for medication; large-vessel disease; abnormal ankle-brachial pressure index; metabolic syndrome	Na intake adjusted for co-morbidities, BMI (obesity), and tobacco and alcohol use
		Weaknesses
		Urine specimens collected after acute stroke may be influenced by medications and acute disease and not reflect usual intake
		Small sample with few outcome events
Strengths	*Strengths*	*Strengths*
1,550-9,400 mg/d	Adjustments for prior stroke or MI, creatinine, BMI, hypertension, DM, atrial fibrillation, smoking, LDL, HDL, treatment allocation (with ramipril, telmitarsan, or both, statins, beta blockers, diuretics, calcium antagonist, antithrombotic therapy), fruit and vegetable consumption, level of exercise, UK excretion, baseline BP, changes in systolic BP from baseline to last follow-up	Repeated excluding first year of follow-up
Weaknesses		Multiple outcomes
Estimated mean Na intake of 4,770 mg/d is much higher than general U.S. population		Detailed data on covariates
		Conducted supplementary analyses to assess impact of including variables that may be in causal pathway, to explore reverse causation, and to look at differential impacts among subgroups
		Weaknesses
	Analysis conducted with and without adjusting for BP	Cutoffs based on examination of data
		Possible reverse causation due to underlying disease

group analyses found no effect of covariates on the associations between sodium excretion and nonfatal CVD events.

Interactions The authors reported HRs to be similar when data were adjusted for blood pressure. Unpublished data sent to the committee indicate no evidence of interactions by prevalence of diabetes.

Studies on Populations with Prehypertension

Cook et al. (2007)

Population size and characteristics Cook et al. (2007) obtained follow-up data from a subset of 2,415 prehypertensive participants included in a previous randomized comparison of 744 and 2,382 participants from two randomized controlled trials (RCTs), the Trials of Hypertension Prevention (TOHP) I and II, respectively, with mortality follow-up on all participants (3,126 total participants and 2,415 responders). (The original studies were conducted to evaluate the effect of dietary sodium on blood pressure, rather than on health outcomes—see Chapter 3.) Despite the original randomized trial design, those remaining in the follow-up of the intervention group were older than those in the follow-up of the control group: 43.4 (±6.6) vs. 42.6 (±6.5) years (p=0.074) in TOHP I and 43.9 (±6.2) vs. 43.3 (±6.1) years (p=0.015) in TOHP II.

Study design, purpose, and length This prospective cohort study was an observational follow-up conducted about 10 years after the end of TOHP I and 5 years after the end of TOHP II. The purpose of the study was to examine the effects of reduced sodium intake on CVD events 10-15 years following the original trial.

Sodium intake measure and method Sodium intake was assessed by repeated measures of 24-hour urine collections over an 18-month period. Additional information was collected on self-reported sodium intake on a final follow-up questionnaire and included participants' preference for salty and low-sodium foods, usual use of low-sodium products, whether they read food labels for sodium, or whether they tracked daily sodium intake to assess long-term patterns of sodium use.

Range of intake, reference, and adjustments Baseline sodium in the intervention group in TOHP I was 154.5 (±59.9) mmol (3,556 [±1,378] mg)/24 hours and in TOHP II was 182.9 (±78.4) mmol (4,207 [±1,803] mg)/24 hours, with similar levels in the control groups. Over the course of the study, the intervention group experienced a sodium reduction of 55.2

mmol (1,270 mg)/24 hours in TOHP I and 42.5 mmol (978 mg)/24 hours in TOHP II, while the control group experienced a reduction of 11.3 mmol (260 mg)/24 hours in TOHP I and 9.8 mmol (225 mg)/24 hours in TOHP II.

Outcome measure, confounders, and adjustments The primary outcome measure was CVD, which included MI, stroke, coronary artery bypass graft, percutaneous transluminal coronary angioplasty, or death with a cardiovascular cause. Adjustments included clinic, age, race, sex, and weight loss intervention assignment. Adjustments in additional analyses included baseline weight and sodium excretion.

Direction and significance of effect In unadjusted analysis, the low-sodium intervention had somewhat lower CVD event risk compared to the control group (with a nonsignificant p=0.21 in analyses stratified by study). However, in models adjusted for trial, clinic, age, race, and sex, the intervention group had 25 percent lower risk of nonfatal CVD compared to control groups (RR=0.75 [CI: 0.57, 0.99] p=0.04), an association that was of borderline statistical significance. Further adjustment for baseline sodium excretion and body weight found a 30 percent lower risk (RR=0.70 [CI: 0.53, 0.94] p=0.02). Analyses for total mortality, which was the only outcome on which information was available in all participants, found a nonsignificant 20 percent lower mortality in the sodium reduction group compared to the usual sodium intake group (RR=0.80 [CI: 0.51, 1.26] p=0.34).

Interactions Although data were not reported, the authors found no evidence of interactions by age or race.

Cook et al. (2009)

Population size and characteristics Cook et al. (2009) also evaluated a slightly smaller number (2,275) of the prehypertensive participants who were in the control arms of TOHP I and II (37 of whom participated in both trials).

Study design, purpose, and length This prospective cohort study examined relationships between sodium and potassium intake and sodium-to-potassium ratio with CVD events among participants from TOHP I and TOHP II over 10-15 years of follow-up.

Sodium intake measure and method Sodium and potassium intake were determined from repeated measures of 24-hour urinary sodium excretion collected at baseline plus scheduled intervals in the TOHP I (five to seven collections over 18 months) and TOHP II (three to five collections over 3-4 years) intervention trials, with a mean of 4.8 measures for all participants.

Range of intake, reference, and adjustments Sodium was determined as mean urinary excretion levels. Mean baseline sodium excretion for all participants (TOHP I and II) was 176.1 mmol (4,050 mg)/24 hours for men and 138.3 mmol (3,105 mg)/24 hours for women. The overall median urinary sodium excretion collected over 18 months for all participants was 158 mmol (3,634 mg)/24 hours (interquartile range, 127-194 mmol [2,921-4,462 mg]/24 hours). For men, the median excretion was 171 mmol (3,933 mg)/24 hours, and for women, it was 134 mmol (3,082 mg)/24 hours.

Outcome measure, confounders, and adjustments Outcome measures (CVD events) and adjustments were as described for Cook et al. (2007).

Direction and significance of effect Analyses for a linear effect of sodium on CVD outcomes found a nonsignificant 25 percent increase risk of CVD associated with a 100 mmol (2,300 mg)/24-hour higher sodium excretion (RR=1.25 [CI: 0.91, 1.72] p=0.18). This association remained after adjustment for baseline and changes in blood pressure and medication use, which suggests that any effect on outcomes may be independent of effects of blood pressure. After adjustment for potassium excretion, the association of sodium and CVD was strengthened and rendered borderline statistically significant. Each 100 mmol (2,300 mg)/24-hour higher sodium excretion was associated with a 42 percent higher risk of CVD events (RR=1.42 [CI: 0.99, 2.04] p=0.05). Across sex-specific quartiles of sodium excretion, using lowest quartile as reference, no significant trend was detected for CVD risk (from lowest to highest, RR=1.00, 0.99 [CI: 0.62, 1.58], 1.16 [CI: 0.73, 1.84], and 1.20 [CI: 0.73, 1.97] p for trend=0.38).

Additional Studies with Analysis of Statistical Interactions by Blood Pressure

Several other studies discussed in this chapter analyzed data on health outcomes by blood pressure and found no statistical interactions (Cohen et al., 2006, 2008; Gardener et al., 2012; O'Donnell et al., 2011; Yang et al., 2011).

Studies on Populations with Diabetes

Ekinci et al. (2011)

Population size and characteristics Ekinci et al. (2011) examined participants with type 2 diabetes (n=638; 56 percent male), who were attending a diabetes clinic in Australia. The mean age of all participants was 64 years and mean duration of diabetes was 11 years. Forty-seven percent of the participants were obese. During follow-up, all patients continued their standard medical care, including antihypertensive, lipid lowering, and anti-diabetic medications. Patients with type 1 diabetes or diabetes secondary to medication or pancreatitis were excluded from the study.

Study design, purpose, and length This prospective cohort study was carried out to examine associations between dietary sodium intake and all-cause and CVD mortality in patients with type 2 diabetes.

Sodium intake measure and method Sodium intake was estimated from 24-hour urine collection. Patients were given general dietary advice as part of their routine care by a dietitian, but follow-up urinalysis for sodium was not performed.

Range of intake, reference, and adjustments Participants were stratified into tertiles: <3,450, 3,540-4,784, and ≥4,785 mg sodium per day. Those in the lowest sodium tertile were older, less likely to be on medication, and had lower estimated glomerular filtration rate (eGFR).

Outcome measure, confounders, and adjustments The primary outcome was death from any cause. Cardiovascular mortality also was included as an outcome. Adjustment was made for age, sex, duration of diabetes, atrial fibrillation, the presence and severity of CKD in the analyses.

Direction and significance of effect The hazard ratios for all-cause mortality and CVD mortality for each 100 mmol per day (2,300 mg per day) higher sodium intake were 0.72 ([CI: 0.55, 0.94] p=0.017) and 0.65 ([CI: 0.44, 0.95] p=0.026), respectively.

Tikellis et al. (2013)

Population size and characteristics Tikellis et al. (2013)[1] analyzed data from the Finnish Diabetic Nephropathy (FinnDiane) study. This study included

---

[1] Published online, December 2012.

TABLE 4-5 Weaknesses and Strengths of Population Studies and Methods of Studies on Cardiovascular Health Outcomes. CVD Outcomes in Populations with Hypertension and Prehypertension*

Study	Sample Size and Population	Study Design and Length	Method to Assess Sodium Intake
Cook et al., 2007	*Strengths*	*Strengths*	*Strengths*
	Good generalizability to the general population or subgroups of interest (prehypertensives, U.S.)	Follow-up of 2 RCTs Randomized lifestyle Na reduction intervention	Multiple UNa collections
Cook et al., 2009	*Strengths*	*Strengths*	*Strengths*
	Good generalizability to the general population or subgroups of interest (prehypertensives, U.S.)	Prospective cohort	Average of 5-7 24-h urine collections taken over 1.5-3 y

* Six studies analyzed the data on health outcomes by prehypertension prevalence or blood pressure and found no interactions (Cohen et al., 2006, 2008; Cook et al., 2009; Gardener et al., 2012; O'Donnell et al., 2011; Yang et al., 2011).
NOTES: Sodium intake presented as mmol in a study was converted to mg using 23 mg/mmol. CVD, cardiovascular disease; d, day; h, hour; mg, milligrams; Na, sodium; RCT, randomized controlled trial; TOHP, Trials of Hypertension Prevention; UNa, urinary sodium; y, year.

participants with type 1 diabetes but excluded those with preexisting CVD as well as those with ESRD. The cohort included 2,648 adults 18 years of age and older, with a mean age of 38 years.

Study design, purpose, and length This prospective cohort study examined associations between sodium intake and CVD and mortality outcomes in adults with type 1 diabetes for a median of 10 years.

Sodium intake measure and method Sodium intake was estimated from one 24-hour urinary sodium collected at baseline.

Range of intake, reference, and adjustments Urinary sodium excretion was categorized into quartiles. The second and third quartiles were collapsed

Sodium Intake Levels or Intake Ranges	Adjustment for Confounders	Other
Strengths	*Strengths*	*Strengths*
Net Na reduction TOHP I: 1,014 mg/24h (from average 3,592 mg/d)	Adjustments for trial, clinic, weight loss intervention	Randomized Na reduction intervention reduces confounding by other factors
TOHP II: 755 mg/24 h (from average 4,250 mg/d)		
Strengths	*Strengths*	*Strengths*
Mean UNa excretion	Adjustments for clinic, treatment assignment, education status, baseline weight, alcohol use, smoking, exercise, family history of CVD, changes in weight, smoking, exercise, K	Follow-up for 10-15 y following Na assessment
Men: 4,050 mg/d		
Women: 3,181 mg/d		

for analysis so that the categories used for analysis were <102 (<2,346), 102-187 (2,346-4,301), or >187 (>4,301) mmol (mg) per day. Adjustments were made for parameters associated with daily urinary sodium excretion.

Outcome measure, confounders, and adjustments Outcome measures were CVD mortality or all-cause mortality. CVD analyses were performed using spline analysis with a single knot at 102 mmol (2,346 mg) per day for CVD incidence and at 141 mmol (3,243 mg) per day for mortality. Adjustments were made for age, sex, glycemic control, presence and severity of CKD, and total cholesterol and triglycerides.

Direction and significance of effect Adjusted multivariate regression analysis found urinary sodium excretion was associated with incident CVD, with

TABLE 4-6 Weaknesses and Strengths of Population Studies and Methods of Studies on Cardiovascular Health Outcomes: CVD Outcomes in Populations with Diabetes*

Study	Sample Size and Population	Study Design and Length	Method to Assess Sodium Intake
Ekinci et al., 2011	*Strengths*	*Strengths*	*Strengths*
	Good generaliza-bility to subgroups of interest (diabetes, Australia)	Prospective cohort	Na estimate based on mean of 1-5 24-h urine collections
			Weaknesses No assessment of dietary intake and thus no ability to adjust for calories
Tikellis et al., 2013	*Strengths*	*Strengths*	*Weaknesses*
		Prospective cohort	Only one assessment of 24-h urine at baseline
			No assessment of adequacy of 24-h urine collections

* Two studies analyzed the data on health outcomes by diabetes prevalence and found no interactions (Cohen et al., 2006; O'Donnell et al., 2011).
NOTES: Sodium intake presented as mmol in a study was converted to mg using 23 mg/mmol. ACE, angiotensin-converting enzyme; BP, blood pressure; CKD, chronic kidney disease; CVD, cardiovascular disease; d, day; eGFR, estimated glomerular filtration rate; h, hour; mg, milligram; Na, sodium; UNa, urinary sodium.

increased risk at both the highest and lowest urine sodium excretion levels. When analyzed as independent outcomes, no significant associations were found between urinary sodium excretion and new CVD or stroke after adjustment for other risk factors.

Additional Studies with Analysis of Data by Diabetes Prevalence

Two other studies discussed in this chapter analyzed the data on health outcomes by diabetes prevalence and found no interaction (Cohen et al., 2006; O'Donnell et al., 2011).

Sodium Intake Levels or Intake Ranges	Adjustment for Confounders	Other
Strengths	*Strengths*	*Weaknesses*
<3,450-4,784 mg/d	Adjustments for duration of diabetes, atrial fibrillation, presence/ severity of CKD	Individuals in lower Na tertiles were older, with lower eGFR and less likely to be on ACE inhibitor
		Na intake adjusted for systolic BP (which had inverse effect on mortality)
		Potential for reverse causation
Strengths	*Strengths*	*Weaknesses*
<2,346-4,301 mg/d	Adjustment for duration of diabetes, presence/severity of CKD, presence of established CVD, systolic BP	Risk among those with extremely low levels of UNa likely reflective of poor health
		Absolute levels of Na unclear in analysis of CVD

Studies in Populations with Congestive Heart Failure

The committee also reviewed evidence on the association of sodium intake with CHF. Its summary of the evidence for CHF is shown in Tables 4-7 and 4-8.

Arcand et al. (2011)

Population size and characteristics Arcand et al. (2011) examined data from 123 New York Heart Association (NYHA)[2] Class I/II and class III/IV medically stable CHF patients enrolled in multidisciplinary CHF programs in two tertiary care hospitals in Canada. Patients were between 18 and 85 years of age, had a left ventricular ejection fraction (LVEF) of <35 percent, were medically stable for at least 3 months, and were on standard medical therapy. Patients were excluded if they had significant renal dysfunction or cardiac cachexia. All patients were consuming a self-selected diet.

Study design, purpose, and length This small prospective cohort study followed participants for 3 years to determine whether a high sodium intake is related to acute decompensated heart failure (ADHF) in ambulatory patients.

Sodium intake measure and method Sodium intake was measured using two 3-day food records: at baseline and after 6-12 weeks. Intake estimates were verified by 24-hour urine analysis in a subset of patients.

Range of intake, reference, and adjustments Calorie-adjusted sodium intakes were pooled as tertiles with cut-points at 1,900 mg per day and 2,700 mg per day.

Outcome measure, confounders, and adjustments The primary outcome measure was ADHF. Secondary outcomes were all-cause hospitalization and death or transplantation. Adjustments were made for age, sex, energy intake, LVEF, beta blockers, furosemide, and BMI.

Direction and significance of effect High sodium intake levels (≥2,800 mg per day) were significantly associated with ADHF (HR=2.55 [CI: 1.61, 4.04] p=0.001), all-cause hospitalization (HR=1.39 [CI: 1.06, 1.83]), and mortality (HR=3.54 [CI: 1.46, 8.62] p=0.005).

---

[2] The NYHA Functional Classification system is used to classify heart failure based on severity of symptoms and how the person feels during physical activity. Class I patients have cardiac disease but no limitation of physical activity. Class II patients experience slight limitations in physical activity (e.g., fatigue and palpitation). Patients classified as Class III experience limitation in less than ordinary physical activity. Class IV patients are unable to participate in physical activity without discomfort and may experience heart failure symptoms at rest (AHA, 2011).

Lennie et al. (2011)

Population size and characteristics Lennie et al. (2011) evaluated a cohort of 302 patients with CHF NYHA Class I to IV recruited from six large community hospitals or medical centers in Georgia, Indiana, Kentucky, and Ohio. Patients were eligible if they had a confirmed diagnosis of chronic CHF with reduced LVEF, had been on medication for at least 3 months, and could read and speak English. Those who were referred for transplantation, had a history of acute MI, valvular heart disease, peripartum heart failure, myocarditis, inflammatory disease, ESRD, or coexisting terminal illness were excluded.

Study design, purpose, and length This small prospective cohort study examined differences in cardiac event-free survival between patients with sodium intake either above or below 3,000 mg per day over a period of 12 months.

Sodium intake measure and method Sodium intake was estimated from a single 24-hour urine collection.

Range of intake, reference, and adjustments The analysis was conducted using 3,000 mg urinary sodium (130 mmol) as the only cut-point for dietary salt intake. Patients were divided into two groups using the 3,000-mg cut-point and stratified by NYHA Class I/II vs. Class III/IV. Adjustment was made for BMI.

Outcome measure, confounders, and adjustments The primary outcomes of the study were the composite endpoint of time to first event for emergency department or hospital admission for CHF or other cardiac-related cause, and all-cause mortality. This model adjusted for age, sex, etiology of CHF, BMI, LVEF, and total comorbidity score.

Direction and significance of effect Results for event-free survival at a urinary sodium of ≥3,000 mg per day varied by the severity of patient symptoms. Among participants stratified into NYHA Class I/II, sodium intake greater than 3,000 mg per day was correlated with a lower disease incidence compared to those with a sodium intake less than 3,000 mg per day (HR=0.44 [CI: 0.20, 0.97] p=0.40). Conversely, participants stratified into NYHA Class III/IV and a sodium intake greater than 3,000 mg per day had a higher disease incidence than those with sodium intakes less than 3,000 mg per day (HR=2.54 [CI: 1.10, 5.84] p=0.028).

Parrinello et al. (2009)

Population size and characteristics Parrinello et al. (2009) examined 173 previously hospitalized patients who had a recent event of decompensated CHF in Italy. Characteristics for inclusion were heart failure consistent with the definition of NYHA functional classification of CHF, uncompensated CHF, or Class IV that was unresponsive to treatment. Participants with cerebral vascular disease, dementia, cancer, uncompensated diabetes, and severe hepatic disease were excluded.

Study design, purpose, and length This RCT was designed to evaluate the effects of dietary sodium restriction at two levels on neurohormonal and cytokine activation and on clinical outcomes over a period of 12 months.

Sodium intake measure and method The treatment groups received one of two levels of a prescribed sodium-restricted diet. Both treatment groups included daily intake of 1,000 ml of fluid and 125-250 mg furosemide twice daily. No follow-up sodium intake assessment was performed.

Range of intake, reference, and adjustments Dietary sodium levels were controlled at either 120 or 80 mmol (2,760 or 1,840 mg) for the modestly restricted and restricted diets, respectively. Adherence to the diet was confirmed by 24-hour urine analysis at baseline, 6 months, and 12 months.

Outcome measure, confounders, and adjustments The primary outcome measures were neurohormonal markers, cytokine levels, hospital readmission, and mortality.

Direction and significance of effect During the 12 months of follow-up, participants receiving the restricted sodium diet had a greater number of hospital readmissions (adjusted risk reduction [ARR]=14.2 [CI: 5.65, 22.7] $p<0.005$) and higher mortality (ARR=14.2 [CI: 5.65, 22.7] $p<0.005$) compared to those on the modestly restricted diet.

Paterna et al. (2008)

Population size and characteristics Paterna et al. (2008) examined 232 Italian patients 53-83 years of age and classified as NYHA Class II and recently hospitalized for decompensated heart failure. Study participants with cerebral vascular disease, dementia, cancer, uncompensated diabetes, and severe hepatic disease were excluded. Eligible patients were those unresponsive to treatment regimens with high doses of oral furosemide up to 250-500 mg per day and/or combinations of diuretics (thiazide, loop diuretic, and spironolactone), angiotensin-converting enzyme (ACE) inhibitors (captopril; 75-150 mg per day), digitalis, beta blockers and nitrates, and who also had aggressive co-treatment with high doses of diuretics (furosemide [250 mg twice daily] and severe fluid restriction [1,000 ml per day]).

Study design, purpose, and length This RCT evaluated the effects of two different sodium levels on risk of hospital readmission. Study participants were evaluated by two physicians blinded to the protocol every week for the first month, then every 2 weeks for the next 2 months, and then every month for the reminder of the study period (6 months).

Sodium intake measure and method Sodium intake levels were prescribed to participants in treatment groups.

Range of intake, reference, and adjustments Participants in treatment groups were randomized to a diet with low (80 mmol per day [1,840 mg per day]) or normal (120 mmol per day [2,760 mg per day]) sodium for 180 days; adherence to the diet prescribed by dietitians was ascertained every week.

Outcome measure, confounders, and adjustments The outcome measure was hospital readmission after having been hospitalized for CHF and then discharged within the previous 30 days.

Direction and significance of effect The lower sodium intake group experienced a significantly higher number of hospital readmissions compared to the normal sodium intake group (absolute risk reduction=18.69% [CI: 9.29, 28.08] $p<0.05$) and a higher but not significantly higher mortality compared to the normal sodium intake group (absolute risk reduction=8.07% [CI: 0.71, 15.43]).

Paterna et al. (2009)

Population size and characteristics Paterna et al. (2009) examined 410 (205 intervention and 205 control) recently hospitalized patients with decompensated CHF NYHA Class II-IV who were 55-83 years of age.

Study design, purpose, and length A 2×2×2 factorial double-blinded RCT was used to assess the effect of dietary sodium intake in combination with a diuretic and fluid regimen on risk of mortality.

Sodium intake measure and method Sodium intake ranges for participants were estimated based on food diaries.

Range of intake, reference, and adjustments Sodium intake ranges were categorized into 120 mmol (2,760 mg) or 80 mmol (1,840 mg) per day.

Outcome measure, confounders, and adjustments The primary outcome measures were hospital readmission and mortality. The independent variables were fluid intake limits (1,000 and 2,000 ml per day), sodium intake (120 and 80 mmol per day [2,760 and 1,840 mg per day]), and furosemide treatment (500 and 250 mg per day). Study participants were evaluated every week for the first month and then every 2 weeks for the next 2 months and then every month for the remainder of the study period (6 months).

Direction and significance of effect A significant association was found between the low sodium intake (80 mmol per day [1,840 mg per day]) and hospital readmissions (odds ratio [OR]=2.46 [CI: 1.84, 3.29] $p<0.0001$). The group with normal sodium diet also had fewer deaths compared to all groups receiving a low-sodium diet combined. The effect of low sodium intake on health outcomes within subgroups by dosage of furosemide and

TABLE 4-7 Weaknesses and Strengths of Population Studies and Methods of Studies on Cardiovascular Health Outcomes. Observational Trials: CVD Outcomes in Populations with CHF

Study	Sample Size and Population	Study Design	Method to Assess Sodium Intake
Arcand et al., 2011	*Strengths*	*Strengths*	*Strengths*
	Good generalizability to subgroups of interest (congestive heart failure, U.S.)	Prospective cohort	Two 3-d food records validated with 2 urine collections
Lennie et al., 2011	*Strengths*	*Strengths*	*Strengths*
	Good generalizability to subgroups of interest (congestive heart failure, U.S.)	Prospective cohort	24-h UNa collection

NOTES: Sodium intake presented as mmol in a study was converted to mg using 23 mg/mmol. BMI, body mass index; CHF, congestive heart failure; CVD, cardiovascular disease; d, day; LVEF, left ventricular ejection fraction; mg, milligram; Na, sodium; UNa, urinary sodium.

fluid restriction could not be examined due to the small number of participants in the subgroups.

SYNTHESIS OF THE EVIDENCE

Methodological Approach

A number of factors influenced the committee's assessment of the evidence reviewed. These included the variability in methodological approaches
used to evaluate relationships between sodium intake and risk of health
outcomes, study design, limitations in the quantitative measures of both
dietary intake and urinary excretion of sodium, confounder adjustment,
and the number of relevant studies available. The committee considered
studies that determined sodium intake levels through multiple high-quality
24-hour urine collections to be the best design. In addition to inconsistencies in sodium intake measures, methodological flaws included the possibility of confounding, measurement error (e.g., systematic underestimates of
sodium intake in those at highest risk of outcomes), and reverse causation
(e.g., individuals with existing underlying disease that leads to low sodium
intake and eventually the outcome of interest).

Sodium Intake Levels or Intake Ranges	Adjustment for Confounders	Other
Strengths	*Strengths*	*Weaknesses*
Mean lowest-highest tertile: 1,400-3,800 mg/d	Adjustments for caloric intake, LVEF, BMI; furosemide use, use of beta blockers	Main outcome (events in upper tertile) not predefined
Strengths	*Strengths*	*Weaknesses*
Mean=4,100 mg/d	Adjustments for CHF etiology, BMI, ejection fraction, total comorbidity score	Na intake dichotomized at 3,000 mg/d so questions regarding Na intake <1,500 mg/d not addressed

TABLE 4-8 Weaknesses and Strengths of Population Studies and Methods of Studies on Cardiovascular Health Outcomes. Randomized Control Trials: CVD Outcomes in Populations with CHF

Study	Sample Size and Population	Study Design	Method to Assess Sodium Intake
Parrinello et al., 2009	*Weaknesses*	*Strengths*	*Strengths*
	Limited generalizability due to eligibility criteria (unresponsive to treatment) and aggressive co-treatment with high dose of diuretics (furosemide 125 to 250 mg twice/d) and severe fluid restriction (1,000 ml/d)	Randomization by a preliminary computer algorithm *Weaknesses* Lack of cotreatment of ACE inhibitors and beta blocker in protocol Confounded by worsened renal function in low-Na treatment by high dose of diuretics and severe fluid restriction	24-hour UNa

Sodium Intake Levels or Intake Ranges	Blinding	Other
Strengths	*Strengths*	*Weaknesses*
80 vs. 120 mmol/d (1,840 vs. 2,760 mg/d)	Double blind (no details)	Unclear if analysis was intention-to-treat

continued

TABLE 4-8 Continued

Study	Sample Size and Population	Study Design	Method to Assess Sodium Intake
Paterna et al., 2008	*Weaknesses*	*Strengths*	*Strengths*
	Limited generalizability due to eligibility criteria (unresponsive to treatment) and aggressive co-treatment with high dose of diuretics (furosemide 250 to 500 mg twice/d) and severe fluid restriction (1,000 ml/d day)	Randomization was carried out using a preliminary computer algorithm and the assignment of all patients was decided at baseline before performing clinical and laboratory measurements	Multiple written standard diets containing 80 or 120 mmol Na (1,840 or 2,760 mg) prepared by dieticians
		Weaknesses	*Weaknesses*
		Lack of co-treatment of ACE inhibitors and beta blocker in protocol	No UNa measures
		Confounded by worsened renal function in low-Na treatment by high dose of diuretics and severe fluid restriction	
		Short follow-up period (180 d)	

Sodium Intake Levels or Intake Ranges	Blinding	Other
Strengths	*Weaknesses*	*Weaknesses*
80 mmol/d vs. 120 mmol/d (1,840 mg/d vs. 2,760 mg/d)	Unblinded	Unclear if analysis was intention-to-treat

continued

TABLE 4-8 Continued

Study	Sample Size and Population	Study Design	Method to Assess Sodium Intake
Paterna et al., 2009	*Weaknesses*	*Strengths*	*Strengths*
	Limited generalizability due to eligibility criteria	Randomization by a preliminary computer algorithm and the assignment of all patients was decided at baseline before performing clinical and laboratory measurements	Patients received multiple written diets containing 80 or 120 mmol Na (1,840 or 2,760 mg/d) prepared by dieticians
	Limited sample of patients in subgroups. Therefore, the effect of low Na intake on clinical outcomes within subgroups by dosage of furosemide and fluid restriction cannot be examined		*Weaknesses*
			No UNa measures

NOTES: Sodium intake presented as mmol in a study was converted to mg using 23 mg/mmol. ACE, angiotensin-converting enzyme; BMI, body mass index; CHF, congestive heart failure; d, day; h, hour; LVEF, left ventricular ejection fraction; mg, milligram; ml, milliliter; mmol, millimole; Na, sodium; UNa, urinary sodium.

Assessing the impact of sodium intake on health outcomes was further complicated by wide variability in intake ranges among studies. For example, in the studies reviewed, high sodium intake ranged from about 2,700 to more than 10,000 mg per day; the high intake ranges of some studies overlapped with the lower ranges in others. The wide range of typical intakes across various population groups, as well as differences in the methods used to measure dietary sodium among different studies, meant that the committee could not derive a numerical definition for high or low intakes in its findings and conclusions. Rather, it could consider sodium intake levels only within the context of an individual study. Thus, in its findings and conclusions, the committee's description of sodium intake reflects the levels in the ranges described in the evidence reviewed. Likewise, the extreme variability in intake levels among population groups precluded the committee from establishing a "healthy" intake range.

Sodium Intake Levels or Intake Ranges	Blinding	Other
Strengths	*Weaknesses*	
80 mmol/d vs. 120 mmol/d (1,840 mg/d vs. 2,760 mg/d)	Unblinded	

Findings for Cardiovascular Disease, Stroke, and Mortality

The committee reviewed evidence that included a broad range of population groups and methodological approaches to determine relationships between sodium intake and direct measures of disease, specifically CVD, stroke, and mortality, including all-cause mortality. All of the evidence considered was observational, mostly prospective cohort studies that examined associations between sodium intake and risk of adverse health outcomes. The populations studied were disproportionately from outside the United States and many included groups that consumed levels of sodium much higher than 3,400 mg per day, the average amount consumed by U.S. adults. Many studies also focused on populations with hypertension, borderline hypertension, or other preexisting diseases or conditions that put them at risk of developing disease.

General Population

Based on its assessment of the evidence, the committee found, first, the evidence reviewed on specific adverse health outcomes consistently indicates an association in the general population between excessive sodium intakes and increased risk of CVD, particularly for stroke. In particular, data from studies using FFQs generally supported an association between high sodium intake and increased risk of CVD, particularly stroke (Gardener et al., 2012; Nagata et al., 2004; Takachi et al., 2010; Umesawa et al., 2008), although not consistently (Larsson et al., 2008). Several of these studies evaluated populations with sodium intakes much higher than the average U.S. intake of 3,400 mg per day (Nagata et al., 2004; Takachi et al., 2010; Umesawa et al., 2008). Gardener et al. (2012), however, found an effect on stroke with sodium intakes starting at 1,500 mg per day (HR=1.17 per 500 mg increase in sodium). Sodium intake data from FFQs, however, are limited by errors in estimating discretionary sodium intake (salt added in cooking or at the table), which accounts for an estimated 11 percent of sodium intakes (Mattes and Donnelly, 1991).

The committee found, in contrast, that the evidence from the current literature is inconsistent with regard to associations with sodium intakes below 2,300 mg per day, with results ranging from lower, similar, or higher risk of CVD, stroke, or mortality, including all-cause mortality. All of the studies identified have limitations of different types. The evidence in some cases is suggestive, however, of associations between lower sodium intake (below 2,300 mg per day) and potential increased risk of adverse health outcomes, though reverse causation, confounding, and systematic measurement error cannot be ruled out.

For example, studies using data from NHANES III (which is representative of the general U.S. population) used 24-hour recall data to estimate sodium intake, which, as discussed in Chapter 2, could introduce considerable error in the measurement of sodium levels. In addition, such studies showed inconsistent results, depending on the methodological approach. Two studies (Cohen et al., 2006, using NHANES II; Cohen et al., 2008, using NHANES III) found an increased risk of CVD at lower sodium levels, while one study (Yang et al., 2011, also using NHANES III) found a lower risk of all-cause mortality at lower sodium intake levels. These studies, however, are limited by the sodium intake measurement used. In addition, they differ on the corrections made for sodium measurements as well as for calorie intake adjustment. Additionally, Cohen et al. (2006, 2008) did not adjust for within-individual day-to-day variation in intake, whereas that adjustment was made in Yang et al. (2011) As another example, Stolarz-Skryzpek et al. (2011), in a general population, also reported higher CVD outcomes in the lower sodium intake group. However, Stolarz-Skryzpek et al. (2011) were limited by the possibility of unmeasured confounding and undercollection of urine specimens in the lowest sodium tertile.

Population Subgroups

Some studies addressed questions related to associations between sodium intake and health outcomes in population subgroups. The committee evaluated one large (O'Donnell et al., 2011) and six small (Arcand et al., 2011; Dong et al., 2009; Ekinci et al., 2011; Heerspink et al., 2012; Kono et al., 2011; Tikellis et al., 2013) prospective cohort studies in patients with preexisting CHF, stroke, MI, CKD, and diabetes, using various methods of sodium assessment. The outcomes estimates were extremely heterogeneous, with HR values ranging between 0.11 (Dong et al., 2009) and 3.54 (Arcand et al., 2011). Because these populations are typically advised to reduce their sodium intake, reverse causation cannot be ruled out as a factor in the relationship.

In contrast, two related observational studies (Cook et al., 2007, 2009) used three to seven 24-hour urine collections, the best available method, to measure sodium intake levels. Both of these studies were conducted in prehypertensive individuals. Cook et al. (2007), an observational follow-up of the TOHP I and II sodium reduction trials, found a 25 percent reduction in CVD incidence, as well as a nonsignificant 20 percent reduction in total mortality when average levels of sodium intake decreased from approximately 3,600 to 2,300 mg per day in the intervention group in TOHP I and from 4,200 to 3,200 mg per day in TOHP II. Cook et al. (2009), also an observational study that followed participants in the TOHP I and II trials, found a linear increase in the risk of total CVD events with increasing sodium intake levels after adjusting for potassium intake. However, there were relatively few outcomes in the lowest ranges of sodium intake, leading to unstable estimates in those ranges. The committee found that these studies suggest an association between a decrease in CVD event rates and sodium intakes down to 2,300 mg per day, and perhaps below, although based on small numbers.

The committee identified and evaluated three RCTs and two cohort studies that examined associations between sodium intake at low, moderate, and high levels and health outcomes in study participants with CHF at various levels of severity. Although the results from the effects of dietary sodium on outcomes in these participants appear inconsistent, several factors might have contributed to the disparate findings. Three RCTs with a similar sodium reduction protocol from a single site in Italy consistently demonstrate higher adverse events (hospital readmission and mortality) associated with lower-sodium diets. Treatment regimens in the three RCTs (Parrinello et al., 2009; Paterna et al., 2008, 2009) included low rates of beta blocker use and high-dose furosemide diuretic use combined with significant fluid restriction, which does not reflect contemporary U.S. management of patients with CHF.[3] However, the committee could not identify weaknesses in the study designs. Further, the uniformity of the results even under very different basic treatments suggests a need for additional trials to be conducted by other investigators with participants under treatments similar to those used in the United States.

The observational studies may suggest a difference in the effect of lower sodium consumption depending on the degree of compensation of the CHF patients (NYHA I/II vs. III/IV), but this was not observed in the Paterna trials (one trial was done in NYHA II and one in decompensated CHF NYHA Class II to IV). Additional difficulty comparing the trials and the observational studies may arise from the differences in CHF patients after hospital discharge (e.g., the Paterna trials) and stable outpatients with heart failure (e.g., the observational studies in CHF clinics in Lennie et al. [2011] and Arcand et al. [2011]). Lastly, adjustment for different potential confounders may have influenced the outcomes observed in the cohort studies, leading to different interpretations. For example, while Arcand et al. (2011) controlled for caloric intake, none of the studies controlled for education, which has been shown to be associated with better health outcomes and lower sodium intake. In contrast to the Italian studies, which show consistency in the results, the results from Arcand et al. (2011) are inconsistent in that the number of hospital readmissions is largest in the middle sodium intake category (66 percent), although it has the lowest mortality (0 percent).

Importantly, CHF is the only health outcome for which RCTs exist. Therefore, although differences in disease management preclude a definitive assessment of the effects of low sodium intake (i.e., 1,840 mg per day) for CHF patients in the United States, the evidence suggests that low sodium intakes may lead to higher risk of adverse events in mid- to late-stage CHF patients with reduced ejection fraction receiving aggressive therapeutic regimens. In addition, a cohort study in a population including individuals with

---

[3] The Guidelines for Heart Failure patients of the Heart Failure Association of America describes therapies appropriate for the different stages of a patient's health (e.g., decompensated heart failure, reduced ejection fraction, end of life). For patients with reduced ejection fraction (≤40 percent), the guideline recommends ACE inhibitors, ARBs (when ACE inhibitors are not tolerated by patients or in post-MI or chronic heart failure patients), beta blockers, and aldosterone antagonists (unless creatinine is >2.5 mg/dL or serum potassium is >5 mmol/L). In addition, diuretic therapy is recommended in certain patients to restore normal volume. For example, the recommended initial daily dose of furosemide is 20-40 mg once or twice per day for a maximum total daily dose of 600 mg. The guideline states that to minimize fluid retention, sodium intake should be limited to 2,000-3,000 mg per day and fluid should be restricted to less than 2,000 ml.

CVD supports the conclusion reached by the Italian studies. That is, low sodium intake levels might result in increased CHF episodes (O'Donnell et al., 2011). However, O'Donnell et al. (2011) used spot urine collection, and the relationship between low sodium intakes and higher CVD risk appeared limited to the outcome of CVD death and CHF, with no apparent relationship between low sodium and risk of MI or stroke.

Nevertheless, three randomized trials in CHF patients (Parrinello et al., 2009; Paterna et al., 2008, 2009) found that a sodium intake of 1,840 (vs. 2,870) mg per day was associated with increased mortality in this population subgroup. These trials, however, were limited to patients with mid- to late-stage CHF and reduced ejection fraction. In addition, patients were receiving aggressive therapeutic regimens that were very different from current standards of care, and thus, the results may not be generalizable.

Finally, the committee found that overall, the paucity of evidence in the general population and population subgroups strongly points to the need for further research to better define relationships between sodium intake and risk of CVD, stroke, and mortality, particularly at the lowest levels of sodium intake within the U.S. population.

STUDIES ON KIDNEY DISEASE

Two studies explored the relationship between sodium intake and CKD (Heerspink et al., 2012; Thomas et al., 2011). Collectively, the committee found these studies lacking in clarity about the risk of kidney disease progression or ESRD associated with sodium intake. Both of the studies reviewed evaluated populations with diabetes who had macroalbuminuria. These two studies show conflicting results about either benefits or risks associated with sodium intake in diabetic patients with macroalbuminuria. Thomas et al. (2011) demonstrated an inverse association of sodium intake with ESRD risk, which suggests that lower sodium intake may increase ESRD risk. On the other hand, the study by Heerspink et al. (2012) suggests that use of renin-angiotensin-aldosterone system (RAAS)-blocking agents may be more beneficial in patients with low sodium intake. RAAS blockade is the first-line therapy for treatment of diabetic nephropathy (Arauz-Pacheo et al., 2003; NKF, 2007) and has consistently been shown to delay progression of diabetic nephropathy into ESRD. Thus, Heerspink et al. (2012) suggest that sodium restriction may be beneficial rather than harmful in preventing kidney disease progression in kidney disease patients with diabetes and macroalbuminuria.

Overall, the studies published since 2003 reviewed by the committee provide inconsistent data about the relationship of sodium intake levels and kidney disease progression in patients with type 2 diabetes and macroalbuminuria, and some evidence suggests that low sodium intake may be harmful in this population. The committee found no studies published since 2003 evaluating the risk of ESRD among individuals with kidney disease at baseline, who had nondiabetic forms of CKD, or in individuals with diabetes but without macroalbuminuria. Some studies suggest that lower sodium intake may lead to lower proteinuria, and that proteinuria is a strong risk factor for CKD progression (described in Chapter 3). However, recent large-scale clinical trials demonstrate that decrements in proteinuria are not always associated with slower progression of CKD (Mann et al., 2008; Parving et al., 2012). For these reasons, to reach its conclusions, the committee relied primarily on data evaluating relationships between sodium level and CKD progression and dialysis initiation rather than changes in proteinuria.

STUDIES ON METABOLIC SYNDROME, DIABETES, AND GASTRIC CANCER

The committee identified from its literature search two cross-sectional studies that examined associations between sodium intake and risk of metabolic syndrome (Rodrigues et al., 2009; Teramoto et al., 2011). The committee also identified two prospective cohort studies that examined associations between sodium intake and risk of developing diabetes (Hu et al., 2005; Roy and Janal, 2010) and one study that examined the role of genetic polymorphisms linked to sodium intake in risk of diabetes (Daimon et al., 2008). These studies did not meet the committee's criteria for further evaluation of the strengths and weaknesses of the study and its relevance to the committee's task.

The committee identified seven prospective cohort studies (Murata et al., 2010; Peleteiro et al., 2011; Shikata et al., 2006; Sjodahl et al., 2008; Takachi et al., 2010; Tsugane et al., 2004; van den Brandt et al., 2003) and five case-control studies (Lazarević et al., 2011; Lee et al., 2003; Pelucchi et al., 2009; Strumylaite et al., 2006; Zhang and Zhang, 2011) that examined associations between sodium intake and risk of gastric cancer. These studies had a number of limitations, including that the reported intakes of the populations studied were not relevant to intake levels in the United States. Overall, the prospective cohort studies showed conflicting results for risk of gastric cancer and the case-control studies were potentially biased due to recall bias. Another possible effect modifier is infection with *H. pylori*. There is no agreement in the published literature, however, about whether the relationship between infection and sodium intake modifies risk of gastric cancer (Lee et al., 2003; Peleteiro et al., 2011). Taken together, the limitations in these studies precluded further evaluation. Although some evidence suggests that high sodium intake may be associated with increased risk of gastric cancer, the committee found that evidence was not definitive for an effect at low intake ranges.

Details about the design, characteristics, and outcomes for each of these studies are tabulated in Appendix F.

ADDITIONAL HEALTH OUTCOMES

In addition to the more commonly studied health outcomes above, the committee identified studies that examined outcomes related to ascites and reflux (Aanen et al., 2006; Gu et al., 2012; Nilsson et al., 2004), pulmonary function, including asthma and pulmonary hyperresponsiveness (Gotshall et al., 2004; Hirayama et al., 2010; Mickleborough et al., 2005; Sausenthaler et al., 2005), genitourinary symptoms, including kidney stone formation and urinary tract symptoms (Eisner et al., 2009; Maserejian et al., 2009; Meschi et al., 2012; Yun et al., 2010), depression (Song, 2009), and quality of life (Ramirez et al., 2004). The studies identified were inconsistent in methodological approach and results and, for the majority of specific outcomes, only one study was found for a given outcome. For example, although three studies addressed the potential association of sodium dietary intake and stone formation (Eisner et al., 2009; Meschi et al., 2012; Yun et al., 2010), the results were inconsistent. Meschi et al. (2012) found a positive correlation between sodium intake and calcium nephrolithiasis in a retrospective study of Italian women and Yun et al. (2010) found that in a cohort of stone formers, those with hypernatriuresis were more likely to develop stones in 3-year follow-up. The role of sodium intake in the risk of stone formation is not clear, and some authors suggest that while an increase in dietary sodium intake might increase urine calcium, it also might increase urine volume and decrease the urinary supersaturation of calcium oxalate (Eisner et al., 2009). In addition, these studies mirror many of the limitations of other studies reviewed in this chapter, including inconsistent and inadequate sodium intake assessment. Although the evidence was insufficient for the committee to draw conclusions about associations between sodium intake and these health outcomes it recognizes that other studies are ongoing and may be useful in the future.

REFERENCES

Aanen, M. C., A. J. Bredenoord, and A. J. P. M. Smout. 2006. Effect of dietary sodium chloride on gastro-oesophageal reflux: A randomized controlled trial. *Scandinavian Journal of Gastroenterology* 41(10):1141-1146.

AHA (American Heart Association). 2011. *Classes of heart failure.* http://www.heart.org/ HEARTORG/Conditions/HeartFailure/AboutHeartFailure/Classes-of-Heart-Failure_ UCM_306328_Article.jsp (accessed March 15, 2013).

Arauz-Pacheo, C., M. A. Parrott, and P. Raskin. 2003. Treatment of hypertension in adults with diabetes. *Diabetes Care* 26(Suppl 1):S80-S82.

Arcand, J., J. Ivanov, A. Sasson, V. Floras, A. Al-Hesayen, E. R. Azevedo, S. Mak, J. P. Allard, and G. E. Newton. 2011. A high-sodium diet is associated with acute decompensated heart failure in ambulatory heart failure patients: A prospective follow-up study. *American Journal of Clinical Nutrition* 93(2):332-337.

ATBC Cancer Prevention Study Group. 1994. The Alpha-Tocopherol, Beta-Carotene Lung Cancer Prevention Study: Design, methods, participant characteristics, and compliance. *Annals of Epidemiology* 4:1-10.

Cohen, H. W., S. M. Hailpern, J. Fang, and M. H. Alderman. 2006. Sodium intake and mortality in the NHANES II follow-up study. *American Journal of Medicine* 119(3):275. e7-275.e14.

Cohen, H. W., S. M. Hailpern, and M. H. Alderman. 2008. Sodium intake and mortality follow-up in the Third National Health and Nutrition Examination Survey (NHANES III). *Journal of General Internal Medicine* 23(9):1297-1302.

Cook, N. R., J. A. Cutler, E. Obarzanek, J. E. Buring, K. M. Rexrode, S. K. Kumanyika, L. J. Appel, and P. K. Whelton. 2007. Long term effects of dietary sodium reduction on cardiovascular disease outcomes: Observational follow-up of the trials of hypertension prevention (TOHP). *British Medical Journal* 334(7599):885-888.

Cook, N. R., E. Obarzanek, J. A. Cutler, J. E. Buring, K. M. Rexrode, S. K. Kumanyika, L. J. Appel, and P. K. Whelton. 2009. Joint effects of sodium and potassium intake on subsequent cardiovascular disease: The trials of hypertension prevention follow-up study. *Archives of Internal Medicine* 169(1):32-40.

Costa, A. P. R., R. C. S. de Paula, G. F. Carvalho, J. P. Araújo, J. M. Andrade, O. L. R. de Almeida, E. C. de Faria, W. M. Freitas, O. R. Coelho, J. A. F. Ramires, J. C. Quinaglia e Silva, and A. C. Sposito. 2012. High sodium intake adversely affects oxidative-inflammatory response, cardiac remodelling and mortality after myocardial infarction. *Atherosclerosis* 222(1):284-291.

Daimon, M., H. Sato, S. Sasaki, S. Toriyama, M. Emi, M. Muramatsu, S. C. Hunt, P. N. Hopkins, S. Karasawa, K. Wada, Y. Jimbu, W. Kameda, S. Susa, T. Oizumi, A. Fukao, I. Kubota, S. Kawata, and T. Kato. 2008. Salt consumption-dependent association of the GNB3 gene polymorphism with type 2 DM. *Biochemical & Biophysical Research Communications* 374(3):576-580.

Dong, J., Y. Li, Z. Yang, and J. Luo. 2010. Low dietary sodium intake increases the death risk in peritoneal dialysis. *Clinical Journal of the American Society of Nephrology* 5(2):240-247.

Eisner, B. H., M. L. Eisenberg, and M. L. Stoller. 2009. Impact of urine sodium on urine risk factors for calcium oxalate nephrolithiasis. *Journal of Urology* 182(5):2330-2333.

Ekinci, E. I., S. Clarke, M. C. Thomas, J. L. Moran, K. Cheong, R. J. MacIsaac, and G. Jerums. 2011. Dietary salt intake and mortality in patients with type 2 diabetes. *Diabetes Care* 34(3):703-709.

Gardener, H., T. Rundek, C. B. Wright, M. S. V. Elkind, and R. L. Sacco. 2012. Dietary sodium and risk of stroke in the Northern Manhattan Study. *Stroke* 43(5):1200-1205.

Geleijnse, J. M., J. C. M. Witteman, T. Stijnen, M. W. Kloos, A. Hofman, and D. E. Grobbee. 2007. Sodium and potassium intake and risk of cardiovascular events and all-cause mortality: the Rotterdam Study. *European Journal of Epidemiology* 22(11):763-770.

Gotshall, R. W., J. J. Rasmussen, and L. J. Fedorczak. 2004. Effect of one week versus two weeks of dietary NaCl restriction on severity of exercise-induced bronchoconstriction. *Journal of Exercise Physiology Online* 7(1):1-7.

Gu, X. B., X. J. Yang, H. Y. Zhu, and B. Y. Xu. 2012. Effect of a diet with unrestricted sodium on ascites in patients with hepatic cirrhosis. *Gut and Liver* 6(3):355-361.

Heerspink, H. J. L., F. A. Holtkamp, H. H. Parving, G. J. Navis, J. B. Lewis, E. Ritz, P. A. De Graeff, and D. De Zeeuw. 2012. Moderation of dietary sodium potentiates the renal and cardiovascular protective effects of angiotensin receptor blockers. *Kidney International* 82(3):330-337.

HHS and USDA (U.S. Department of Health and Human Services and U.S. Department of Agriculture). 2010a. *Dietary Guidelines for Americans, 2010.* Washington, DC: U.S. Government Printing Office.http://www.cnpp.usda.gov/Publications/DietaryGuidelines/2010/PolicyDoc/qPolicyDoc.pdf (accessed February 4, 2013).

HHS and USDA. 2010b. *Report of the Dietary Guidelines Advisory Committee on the Dietary Guidelines for Americans, 2010, to the Secretary of Agriculture and the Secretary of Health and Human Services. Washington, DC: USDA/ARS.* http://www.cnpp.usda.gov/Publications/DietaryGuidelines/2010/DGAC/Report/2010DGACReport-camera-ready-Jan11-11.pdf (accessed February 1, 2013).

Hirayama, F., A. H. Lee, A. Oura, M. Mori, N. Hiramatsu, and H. Taniguchi. 2010. Dietary intake of six minerals in relation to the risk of chronic obstructive pulmonary disease. *Asia Pacific Journal of Clinical Nutrition* 19(4):572-577.

Hu, G., P. Jousilahti, M. Peltonen, J. Lindstrom, and J. Tuomilehto. 2005. Urinary sodium and potassium excretion and the risk of type 2 diabetes: A prospective study in Finland. *Diabetologia* 48(8):1477-1483.

IOM (Institute of Medicine). 2005. *Dietary reference intakes for water, potassium, sodium, chloride, and sulfate.* Washington, DC: The National Academies Press.

IOM. 2010. *Strategies to reduce sodium intake in the United States.* Washington, DC: The National Academies Press.

Kono, Y., S. Yamada, K. Kamisaka, A. Araki, Y. Fujioka, K. Yasui, Y. Hasegawa, and Y. Koike. 2011. Recurrence risk after noncardioembolic mild ischemic stroke in a Japanese population. *Cerebrovascular Diseases* 31(4):365-372.

Larsson, S. C., M. J. Virtanen, M. Mars, S. Männistö, P. Pietinen, D. Albanes, and J. Virtamo. 2008. Magnesium, calcium, potassium, and sodium intakes and risk of stroke in male smokers. *Archives of Internal Medicine* 168(5):459-465.

Lazarević, K., A. Nagorni, D. Bogdanović, N. Rančić, L. Stošić, and S. Milutinović. 2011. Dietary micronutrients and gastric cancer: Hospital based study. *Central European Journal of Medicine* 6(6):783-787.

Lee, S. A., D. Kang, K. N. Shim, J. W. Choe, W. S. Hong, and H. Choi. 2003. Effect of diet and *Helicobacter pylori* infection to the risk of early gastric cancer. *Journal of Epidemiology* 13(3):162-168.

Lennie, T. A., E. K. Song, J. R. Wu, M. L. Chung, S. B. Dunbar, S. J. Pressler, and D. K. Moser. 2011. Three gram sodium intake is associated with longer event-free survival only in patients with advanced heart failure. *Journal of Cardiac Failure* 17(4):325-330.

Mann, J. F., R. E. Schmieder, M. McQueen, L. Dyal, H. Schumacher, J. Pogue, X. Wang, A. Maggioni, A. Budaj, S. Chaithiraphan, K. Dickstein, M. Keltai, M. Metsärinne, A. Oto, A. Parkhomenko, L. S. Piegas, T. L. Svendsen, K. K. Teo, and S. Yusuf. 2008. Renal outcomes with telmisartan, ramipril, or both, in people at high vascular risk (the ONTARGET study): A multicentre, randomised, double-blind, controlled trial. *The Lancet* 372(9638):547-553.

Maserejian, N. N., E. L. Giovannucci, and J. B. McKinlay. 2009. Dietary macronutrients, cholesterol, and sodium and lower urinary tract symptoms in men. *European Urology* 55(5):1179-1189.

Mattes, R. D., and D. Donnelly. 1991. Relative contributions of dietary sodium sources. *Journal of the American College of Nutrition* 10(4):383-393.

Meschi, T., A. Nouvenne, A. Ticinesi, B. Prati, A. Guerra, F. Allegri, F. Pigna, L. Soldati, G. Vezzoli, G. Gambaro, F. Lauretani, M. Maggio, and L. Borghi. 2012. Dietary habits in women with recurrent idiopathic calcium nephrolithiasis. *Journal of Translational Medicine* 10(1).

Mickleborough, T. D., M. R. Lindley, and S. Ray. 2005. Dietary salt, airway inflammation, and diffusion capacity in exercise-induced asthma. *Medicine and Science in Sports and Exercise* 37(6):904-914.

Murata, A., Y. Fujino, T. M. Pham, T. Kubo, T. Mizoue, N. Tokui, S. Matsuda, and T. Yoshimura. 2010. Prospective cohort study evaluating the relationship between salted food intake and gastrointestinal tract cancer mortality in Japan. *Asia Pacific Journal of Clinical Nutrition* 19(4):564-571.

Nagata, C., N. Takatsuka, N. Shimizu, and H. Shimizu. 2004. Sodium intake and risk of death from stroke in Japanese men and women. *Stroke* 35(7):1543-1547.

Nilsson, M., R. Johnsen, W. Ye, K. Hveem, and J. Lagergren. 2004. Lifestyle related risk factors in the aetiology of gastrooesophageal reflux. *Gut* 53(12):1730-1735.

NKF (National Kidney Foundation). 2007. *KDOQI clinical practice guidelines and clinical practice recommendations for diabetes and chronic kidney disease—guidelines 3: management of hypertension in diabetes and chronic kidney disease.* http://www.kidney.org/professionals/kdoqi/guideline_diabetes/guide3.htm (accessed April 10, 2013).

O'Donnell, M. J., S. Yusuf, A. Mente, P. Gao, J. F. Mann, K. Teo, M. McQueen, P. Sleight, A. M. Sharma, A. Dans, J. Probstfield, and R. E. Schmieder. 2011. Urinary sodium and potassium excretion and risk of cardiovascular events. *Journal of the American Medical Association* 306(20):2229-2238.

Parrinello, G., P. Di Pasquale, G. Licata, D. Torres, M. Giammanco, S. Fasullo, M. Mezzero, and S. Paterna. 2009. Long-term effects of dietary sodium intake on cytokines and neurohormonal activation in patients with recently compensated congestive heart failure. *Journal of Cardiac Failure* 15(10):864-873.

Parving, H. H., B. M. Brenner, J. J. V. McMurray, D. De Zeeuw, S. M. Haffner, S. D. Solomon, N. Chaturvedi, F. Persson, A. S. Desai, M. Nicolaides, A. Richard, Z. Xiang, P. Brunel, and M. A. Pfeffer. 2012. Cardiorenal end points in a trial of aliskiren for type 2 diabetes. *New England Journal of Medicine* 367(23):2204-2213.

Paterna, S., P. Gaspare, S. Fasullo, F. M. Sarullo, and P. Di Pasquale. 2008. Normal-sodium diet compared with low-sodium diet in compensated congestive heart failure: Is sodium an old enemy or a new friend? *Clinical Science* 114(3):221-230.

Paterna, S., G. Parrinello, S. Cannizzaro, S. Fasullo, D. Torres, F. M. Sarullo, and P. Di Pasquale. 2009. Medium term effects of different dosage of diuretic, sodium, and fluid administration on neurohormonal and clinical outcome in patients with recently compensated heart failure. *American Journal of Cardiology* 103(1):93-102.

Peleteiro, B., C. Lopes, C. Figueiredo, and N. Lunet. 2011. Salt intake and gastric cancer risk according to *Helicobacter pylori* infection, smoking, tumour site and histological type. *British Journal of Cancer* 104(1):198-207.

Pelucchi, C., I. Tramacere, P. Bertuccio, A. Tavani, E. Negri, and C. La Vecchia. 2009. Dietary intake of selected micronutrients and gastric cancer risk: An Italian case-control study. *Annals of Oncology* 20(1):160-165.

Ramirez, E. C., L. C. Martinez, A. O. Tejeda, V. R. Gonzalez, R. N. David, and E. A. Lafuente. 2004. Effects of a nutritional intervention on body composition, clinical status, and quality of life in patients with heart failure. *Nutrition* 20(10):890-895.

Rodrigues, S. L., M. P. Baldo, R. de Sa Cunha, R. V. Andreao, M. Del Carmen Bisi Molina, C. P. Goncalves, E. M. Dantas, and J. G. Mill. 2009. Salt excretion in normotensive individuals with metabolic syndrome: A population-based study. *Hypertension Research—Clinical & Experimental* 32(10):906-910.

Roger, V. L., A. S. Go, D. M. Lloyd-Jones, R. J. Adams, J. D. Berry, T. M. Brown, M. R. Carnethon, S. Dai, G. de Simone, E. S. Ford, C. S. Fox, H. J. Fullerton, C. Gillespie, K. J. Greenlund, S. M. Hailpern, J. A. Heit, P. M. Ho, V. J. Howard, B. M. Kissela, S. J. Kittner, D. T. Lackland, J. H. Lichtman, L. D. Lisabeth, D. M. Makuc, G. M. Marcus, A. Marelli, D. B. Matchar, M. M. McDermott, J. B. Meigs, C. S. Moy, D. Mozaffarian, M. E. Mussolino, G. Nichol, N. P. Paynter, W. D. Rosamond, P. D. Sorlie, R. S. Stafford, T. N. Turan, M. B. Turner, N. D. Wong, and J. Wylie-Rosett. 2011. Heart disease and stroke statistics—2011 update: A report from the American Heart Association. *Circulation* 123(4):e18-e209.

Roy, M. S., and M. N. Janal. 2010. High caloric and sodium intakes as risk factors for progression of retinopathy in type 1 diabetes mellitus. *Archives of Ophthalmology* 128(1):33-39.

Sausenthaler, S., I. Kompauer, S. Brasche, J. Linseisen, and J. Heinrich. 2005. Sodium intake and bronchial hyperresponsiveness in adults. *Respiratory Medicine* 99(7):864-870.

Shikata, K., Y. Kiyohara, M. Kubo, K. Yonemoto, T. Ninomiya, T. Shirota, Y. Tanizaki, Y. Doi, K. Tanaka, Y. Oishi, T. Matsumoto, and M. Iida. 2006. A prospective study of dietary salt intake and gastric cancer incidence in a defined Japanese population: The Hisayama study. *International Journal of Cancer* 119(1):196-201.

Sjödahl, K., C. Jia, L. Vatten, T. Nilsen, K. Hveem, and J. Lagergren. 2008. Salt and gastric adenocarcinoma: A population-based cohort study in Norway. *Cancer Epidemiology Biomarkers and Prevention* 17(8):1997-2001.

Song, E. K. 2009. Adherence to the low-sodium diet plays a role in the interaction between depressive symptoms and prognosis in patients with heart failure. *Journal of Cardiovascular Nursing* 24(4):299-305.

Stolarz-Skrzypek, K., T. Kuznetsova, L. Thijs, V. Tikhonoff, J. Seidlerová, T. Richart, Y. Jin, A. Olszanecka, S. Malyutina, E. Casiglia, J. Filipovský, K. Kawecka-Jaszcz, Y. Nikitin, and J. A. Staessen. 2011. Fatal and nonfatal outcomes, incidence of hypertension, and blood pressure changes in relation to urinary sodium excretion. *Journal of the American Medical Association* 305(17):1777-1785.

Strumylaite, L., J. Zickute, J. Dudzevicius, and L. Dregval. 2006. Salt-preserved foods and risk of gastric cancer. *Medicina* 42(2):164-170.

Takachi, R., M. Inoue, T. Shimazu, S. Sasazuki, J. Ishihara, N. Sawada, T. Yamaji, M. Iwasaki, H. Iso, Y. Tsubono, and S. Tsugane. 2010. Consumption of sodium and salted foods in relation to cancer and cardiovascular disease: The Japan Public Health Center-based prospective study. *American Journal of Clinical Nutrition* 91(2):456-464.

Teramoto, T., R. Kawamori, S. Miyazaki, and S. Teramukai. 2011. Sodium intake in men and potassium intake in women determine the prevalence of metabolic syndrome in Japanese hypertensive patients: OMEGA Study. *Hypertension Research* 34(8):957-962.

Thomas, M. C., J. Moran, C. Forsblom, V. Harjutsalo, L. Thorn, A. Ahola, J. Waden, N. Tolonen, M. Saraheima, D. Gordin, and P. H. Groop. 2011. The association between dietary sodium intake, ESRD, and all-cause mortality in patients with type 1 diabetes. *Diabetes Care* 34(4):861-866.

Tikellis, C., R. J. Pickering, D. Tsorotes, V. Harjutsalo, L. Thorn, A. Ahola, J. Waden, N. Tolonen, M. Saraheimo, D. Gordin, C. Forsblom, P. H. Groop, M. E. Cooper, J. Moran, and M. C. Thomas. 2013. Association of dietary sodium intake with atherogenesis in experimental diabetes and with cardiovascular disease in patients with type 1 diabetes. *Clinical Science* 124(10):617-626.

Tooze, J. A., D. Midthune, K. W. Dodd, L. S. Freedman, S. M. Krebs-Smith, A. F. Subar, P. M. Guenther, R. J. Carroll, and V. Kipnis. 2006. A new statistical method for estimating the usual intake of episodically consumed foods with application to their distribution. *Journal of the American Dietetic Association* 106(10):1575-1587.

Tsugane, S., S. Sasazuki, M. Kobayashi, and S. Sasaki. 2004. Salt and salted food intake and subsequent risk of gastric cancer among middle-aged Japanese men and women. *British Journal of Cancer* 90(1):128-134.

Umesawa, M., H. Iso, C. Date, A. Yamamoto, H. Toyoshima, Y. Watanabe, S. Kikuchi, A. Koizumi, T. Kondo, Y. Inaba, N. Tanabe, and A. Tamakoshi. 2008. Relations between dietary sodium and potassium intakes and mortality from cardiovascular disease: The Japan Collaborative Cohort study for evaluation of cancer risks. *American Journal of Clinical Nutrition* 88(1):195-202.

van den Brandt, P. A., A. A. M. Botterweck, and R. A. Goldbohm. 2003. Salt intake, cured meat consumption, refrigerator use and stomach cancer incidence: A prospective cohort study (Netherlands). *Cancer Causes and Control* 14(5):427-438.

Yang, Q., T. Liu, E. V. Kuklina, W. D. Flanders, Y. Hong, C. Gillespie, M. H. Chang, M. Gwinn, N. Dowling, M. J. Khoury, and F. B. Hu. 2011. Sodium and potassium intake and mortality among US adults: Prospective data from the Third National Health and Nutrition Examination Survey. *Archives of Internal Medicine* 171(13):1183-1191.

Yun, S. J., Y. S. Ha, W. T. Kim, Y. J. Kim, S. C. Lee, and W. J. Kim. 2010. Sodium restriction as initial conservative treatment for urinary stone disease. *Journal of Urology* 184(4):1372-1376.

Zhang, Z., and X. Zhang. 2011. Salt taste preference, sodium intake and gastric cancer in China. *Asian Pacific Journal of Cancer Prevention* 12(5):1207-1210.

5

Findings and Conclusions

INTRODUCTION

Dietary intake of sodium among the general adult U.S. population averages 3,400 mg daily, while federal nutrition policy guidance, the *Dietary Guidelines for Americans 2010* (USDA and HHS, 2010a), recommends sodium intake be reduced to less than 2,300 mg daily and to 1,500 mg daily for African Americans, individuals with hypertension, diabetes, or chronic kidney disease (CKD), and individuals 51 years of age and older. As discussed in Chapter 3, current recommendations for dietary sodium intake are based importantly on the use of blood pressure as a surrogate marker for cardiovascular disease (CVD) outcomes (IOM, 2005; USDA and HHS, 2010b). In particular, in establishing Dietary Reference Intake values for sodium, the Panel on Dietary Reference Intakes for Electrolytes and Water (IOM, 2005) found insufficient evidence to derive an Estimated Average Requirement and calculate a Recommended Dietary Allowance, and thus Adequate Intakes (AIs) were set instead, as an amount needed to achieve a diet that is adequate in other essential nutrients and covers sweat losses. In addition, a Tolerable Upper Intake Level (UL) was established for sodium based on evidence showing associations between high sodium intake and risk of high blood pressure.[1] Thus, the Institute of Medicine (IOM) panel's recommendation was based on an AI for sodium intake of 1,500 mg per

---

[1] Applies to individuals aged 14 years and above.

day and projections from effects on blood pressure that dietary sodium intake up to 2,300 mg daily[2] is not likely to cause any harm.

The relationship between other electrolytes and changes in blood pressure remains unresolved. Yet, as noted in the *Report of the Dietary Guidelines Advisory Committee (DGAC)* (USDA and HHS, 2010b), the effects of lowering sodium intake on blood pressure cannot always be disentangled from the effects of total dietary modification. For example, the committee's review revealed that in a number of studies the effects of dietary sodium on CVD outcomes sometimes persisted even after controlling for blood pressure, suggesting that associations between dietary sodium and risk of CVD may be mediated through other dietary factors (e.g., the effects of other electrolytes), or through pathways in addition to blood pressure. Further, older data indicate that some people in the population may be salt sensitive, while others are not, and that blood pressure response to sodium varies widely (see Chapter 3). In this context, new data have raised questions about the health effects of lowering sodium intake on health outcomes.

Thus, in response to its charge, the committee focused its examination of evidence on the associations between dietary sodium intake and direct health outcomes, not on blood pressure as an indirect or intermediate marker of CVD outcomes (see Appendix D). In deriving its findings and conclusions about the evidence for associations between dietary sodium intake and health outcomes, the committee examined the quality as well as the quantity of the evidence. The conclusions and recommendations drawn from the committee's findings are described below.

FINDINGS AND CONCLUSIONS

The committee's assessment of the evidence reviewed was guided by a number of factors. These included the study design, the quantitative measures of dietary sodium intake and confounder adjustment, and the number and consistency of relevant studies available.

From the evidence reviewed on health outcomes, the committee found that a number of the populations evaluated were outside the United States and included groups that consumed mean levels of sodium much higher than 3,400 mg per day, the average amount consumed by adults in the United States (USDA and HHS, 2010b). Thus, the applicability of some of the results to the U.S. population was of concern. For example, in the studies reviewed, "high" sodium intake levels ranged from about 2,700 to more than 10,000 mg per day.

Overall, the committee found both the quantity and quality of relevant studies to be less than optimal. Further, almost all of the evidence on

---

[2] Applies to adults aged 19-50 years.

clinical outcomes identified by the committee was observational, consisting largely of prospective cohort studies. The committee also found important limitations associated with the quantitative measures of sodium intake (see Chapters 2 and 4) and recognized the potential for spurious findings related to incorrect measurement and reverse causality. Specifically, in some studies, low sodium intakes apparently appeared to show an association with risk of disease, when, in fact, the relationship may have been that the disease itself led to low or incomplete measures of sodium among those with pre-existing disease (see Chapter 4).

Assessing the impact of sodium intake on health outcomes also was complicated by variability in the types and quality of measures used, so that measures could not be reliably calibrated across studies. It was the consensus of the committee that the lack of consistency among studies in the methods used for defining sodium intakes at both high and low ends of the range of typical intakes among various population groups precluded deriving a numerical definition for high and low intakes in its findings and conclusions. Rather, it could consider sodium intake levels only within the context of each individual study. Likewise, the extreme variability in intake levels between and among population groups precluded the committee from establishing a "healthy" intake range.

Overarching Findings

Recognizing the limitations of the available evidence, the committee found no consistent evidence to support an association between sodium intake and either a beneficial or adverse effect on health outcomes other than CVD outcomes (including stroke and CVD mortality) and all-cause mortality. Some evidence suggested that decreasing sodium intake could possibly reduce the risk of gastric cancer. However, the evidence was too limited to conclude the converse—that higher sodium intake could possibly increase the risk of gastric cancer. Interpreting these findings was particularly challenging because most gastric cancer studies were conducted outside the United States in populations consuming much higher levels of sodium than those consumed in this country. Thus, the committee focused its findings and conclusions on evidence for associations between sodium intake and risk of CVD-related events and mortality.

Findings and Conclusions for Cardiovascular
Disease, Stroke, and Mortality

General U.S. Population

Finding 1: The committee found that the results from studies linking dietary sodium intake with direct health outcomes were highly variable in methodological quality, particularly in assessing sodium intake. The range of limitations included over- or underreporting of intakes or incomplete collection of urine samples. In addition, variability in data collection methodologies limited the committee's ability to compare results across studies.

Conclusion 1: Although the reviewed evidence on associations between sodium intake and direct health outcomes has methodological flaws and limitations, the committee concluded that, when considered collectively, it indicates a positive relationship between higher levels of sodium intake and risk of CVD. This evidence is consistent with existing evidence on blood pressure as a surrogate indicator of CVD risk.

Finding 2: The committee found that the evidence from studies on direct health outcomes was insufficient and inconsistent regarding an association between sodium intake below 2,300 mg per day and either benefit or risk of CVD outcomes (including stroke and CVD mortality) or all-cause mortality in the general U.S. population.

Conclusion 2: The committee determined that evidence from studies on direct health outcomes is inconsistent and insufficient to conclude that lowering sodium intakes below 2,300 mg per day either increases or decreases risk of CVD outcomes (including stroke and CVD mortality) or all-cause mortality in the general U.S. population.

Population Subgroups

Finding 1: The committee found that the evidence from multiple randomized controlled trials (RCTs) that were conducted by a single investigative team indicated that low sodium intake (e.g., down to 1,840 mg per day) may lead to greater risk of adverse events in congestive heart failure (CHF) patients with reduced ejection fraction and who are receiving certain aggressive therapeutic regimens. This association also is supported by one observational study where low sodium intake levels in patients with CVD and diabetes were associated with higher risk of CHF events.

Conclusion 1: The committee concluded that the available evidence suggests that low sodium intakes may lead to higher risk of adverse events in mid- to late-stage CHF patients with reduced ejection fraction and who are receiving aggressive therapeutic regimens. Because these therapeutic regimens were very different than current standards of care in the United States, the results may not be generalizable. Similar studies in other settings, and using regimens more closely resembling those in standard U.S. clinical practice are needed.

Finding 2: The committee found that data among prehypertensive participants from two related studies provided some evidence suggesting a continued benefit of lowering sodium intake in these patients down to 2,300 mg per day (and lower, although based on small numbers in the lower range). In contrast, the committee found no evidence for benefit and some evidence suggesting risk of adverse health outcomes associated with sodium intake levels in ranges approximating 1,500 to 2,300 mg per day in other disease-specific population subgroups, specifically those with diabetes, CKD, or preexisting CVD. In addition to inconsistencies in sodium intake measures, methodological flaws included the possibility of confounding and reverse causality. No relevant evidence was found on health outcomes for other population subgroups considered (i.e., persons 51 years of age and older, and African Americans). In studies that explored interactions, race, age, or the presence of hypertension or diabetes did not change the effect of sodium on health outcomes.

Conclusion 2: The committee concluded that, with the exception of the CHF patients described above, the current body of evidence addressing the association between low sodium intake and health outcomes in the population subgroups considered[3] is limited. The evidence available is inconsistent and limited in its approaches to measuring sodium intake. The evidence also is limited by small numbers of health outcomes and the methodological constraints of observational study designs, including the potential for reverse causality and confounding.

The committee further concluded that, while the current literature provides some evidence for adverse health effects of low sodium intake among individuals with diabetes, CKD, or preexisting CVD, the evidence on both benefit and harm is not strong enough to indicate that these subgroups should be treated differently from the general U.S. population. Thus, the committee concluded that the evidence on direct health outcomes does not

---

[3] For example, diabetes, CKD, or preexisting CVD, individuals with hypertension, prehypertension, persons 51 years of age and older, and African Americans.

support recommendations to lower sodium intake within these subgroups to or even below 1,500 mg per day.

Implications for Population-Based Strategies to Gradually Reduce Sodium Intake in the U.S. Population

As noted in Chapter 1, recommendations of the Panel on Dietary Reference Intakes for Electrolytes and Water (IOM, 2005) of an AI for sodium intake of 1,500 mg per day for all individuals 9 years of age up to 51 years of age was set as an amount necessary to achieve an overall diet that provides an adequate intake of other important nutrients and also covers sodium sweat losses. A UL for sodium was set at 2,300 mg per day based on evidence showing associations between high sodium intakes and risk of high blood pressure and consequent risk of CVD, stroke, and mortality.

Given this background, overall, the committee found that the available evidence on associations between sodium intake and direct health outcomes is consistent with population-based efforts to lower excessive dietary sodium intakes, but it is not consistent with previous efforts that encourage lowering of dietary sodium in the general population to 1,500 mg per day. Further, as noted in the DGAC report, population subgroups, including those with diabetes, CKD, or preexisting CVD, individuals with hypertension, prehypertension, persons 51 years of age and older, and African Americans, represent, in aggregate, a majority of the general U.S. population. Thus, when considered in light of the current state of the evidence on associations between sodium intake and direct health outcomes for those subgroups, except when data specifically indicate they are different, there is not sufficient evidence to support treating them differently from the general U.S. population.

The committee was not asked to draw conclusions about a specific target range of dietary sodium for the general population or for population subgroups. However, the committee notes that there are important factors it considered that preclude such a conclusion. For example, one factor that is often discussed in the context of other health-related questions is the challenge of defining specific intake levels when the variables of interest are continuous. That is an especially difficult issue in the present circumstances, where the target intake level could theoretically differ for different large population subgroups.

Other methodological factors that preclude making conclusions about a specific target range for sodium relate to the variability in approaches and study designs in the literature reviewed. Most importantly, quantitative methods for measuring dietary sodium intake have limitations and there are impediments to calibrating those measures across different methodological approaches and study designs. Methodologic problems in assessing sodium intake make this particularly challenging.

FUTURE RESEARCH TO ADDRESS GAPS
IN DATA AND METHODOLOGY

The committee identified a number of data and methods gaps in studies on sodium intake and risk of adverse health outcomes among population groups. Further research in the areas highlighted below would strengthen the evidence base on the association between lower (1,500 to 2,300 mg) levels of sodium and health outcomes in the general population and population subgroups:

1. standardized methodological approaches to measure sodium intake in population groups. Specific examples include standardizing the use of multiple 24-hour urine collections and validating sodium intake estimates with data on urine volume, urine creatinine, and body weight;
2. approaches using dietary sodium intake levels corresponding to levels in current guidelines (i.e., 1,500 to 2,300 mg per day) when examining associations between sodium intake and health outcomes;
3. analyses examining the effects on health outcomes of dietary sodium in combination with other electrolytes, particularly potassium;
4. methods that account for potential confounding factors in dietary studies, including the influence of reported total daily caloric intake on observational associations between sodium and health outcomes, and methods that clarify attributes of individuals with apparently low sodium intake or excretion; and
5. analyses of interactions with antihypertensive medication and blood pressure in studies examining associations between sodium intake and health outcomes.

In addition, the committee identified a need for RCT research, and observational and mechanistic studies, particularly in population subgroups. Examples of such clinical trials include those to examine

1. effects of a range of sodium levels on risk of cardiovascular events, stroke, and mortality among
   a. patients in controlled environments, where randomized trials may be more feasible, such as the elderly in chronic care facilities and other institutionalized individuals; and

   b. individuals as part of natural experiments, such as those in other
      countries where policies affecting sodium consumption are in
      effect;
2. effects of low-sodium diets on adverse events among CHF patients
   receiving therapeutic treatment modalities typically used in the
   United States; and
3. potential beneficial or adverse outcomes of a range of sodium
   intakes among African Americans, adults 51-70 years of age, 70
   years of age and older, and other population subgroups; RCTs may
   be particularly important within higher-risk patient populations,
   where reverse causation is a potential limitation of observational
   studies.

The committee also identified a need for studies to collect and reanalyze

1. data from existing clinical trials that were designed to evaluate
   sodium and health; and
2. data during extended follow-up periods after completion of a clinical trial to identify health outcomes, such as mortality, that could
   manifest later in life and after longer follow-up periods. Such trials
   would not be simple to conduct, however, and careful feasibility
   assessment is needed first.

   In addition to RCT research, mechanistic studies are needed to examine potential physiologic changes associated with lowering sodium intake
and adverse health outcomes. Finally, additional observational research is
needed to examine associations between sodium intake and cancer, especially gastric cancer in the U.S. population, as well as associations between
sodium intake and caloric intake in both short-term and longitudinal
studies.

REFERENCES

IOM (Institute of Medicine). 2005. *Dietary reference intakes for water, potassium, sodium,
   chloride, and sulfate.* Washington, DC: The National Academies Press.
USDA and HHS (U.S. Department of Agriculture and U.S. Department of Health and
   Human Services). 2010a. *Dietary Guidelines for Americans, 2010.* 7th ed. Washing-
   ton, DC: U.S. Government Printing Office. http://www.cnpp.usda.gov/Publications/
   DietaryGuidelines/2010/PolicyDoc/PolicyDoc.pdf (accessed February 4, 2013).
USDA and HHS. 2010b. *Report of the Dietary Guidelines Advisory Committee on the Dietary
   Guidelines for Americans, 2010, to the Secretary of Agriculture and the Secretary of
   Health and Human Services.* Washington, DC: USDA/ARS. http://www.cnpp.usda.gov/
   Publications/DietaryGuidelines/2010/DGAC/Report/2010DGACReport-camera-ready-
   Jan11-11.pdf (accessed February 1, 2013).

Appendix A

Acronyms and Abbreviations

ACE	angiotensin-converting enzyme
ADHF	acute decompensated heart failure
AI	adequate intake
ARB	angiotensin receptor blocker
BMI	body mass index
BP	blood pressure
CDC	Centers for Disease Control and Prevention
CHF	congestive heart failure
CI	confidence interval
CKD	chronic kidney disease
CT	computed tomography
CV	cardiovascular
CVD	cardiovascular disease
d	day
DASH	Dietary Approaches to Stop Hypertension
DGA	*Dietary Guidelines for Americans*
DGAC	Dietary Guidelines Advisory Committee
DM	diabetes mellitus
DRI	Dietary Reference Intake
EAR	Estimated Average Requirement
eGFR	estimated glomerular filtration rate

eNOS	endothelial nitric oxide synthase
ESRD	end-stage renal disease
FFQ	food frequency questionnaire
g	gram
h	hour
HDL	high-density lipoprotein
Hg	mercury
HHS	U.S. Department of Health and Human Services
HR	hazard ratio
IHD	ischemic heart disease
IOM	Institute of Medicine
IS	ischemic stroke
K	potassium
kg	kilogram
KHANES	Korean Health and Nutrition Examination Survey
L	liter
LDL	low-density lipoprotein
LOAEL	lowest observed adverse effect level
LVEF	left ventricular ejection fraction
mg	milligram
MI	myocardial infarction
ml	milliliter
mm	millimeter
mmol	millimole
n	number
Na	sodium
NHANES	National Health and Nutrition Examination Survey
NOAEL	no-observed adverse effect level
ONTARGET	ONgoing Telmisartan Alone in combination with Ramipril Global Endpoint Trial
PRA	plasma renin activity
PURE	Prospective Urban Rural Epidemiology

Q	quartile/quintile
RAAS	renin-angiotensin-aldosterone system
RCT	randomized controlled trial
RDA	Recommended Dietary Allowance
RR	relative risk
SALTURK	Relationship between Hypertension and Salt Intake in Turkish Populations
SD	standard deviation
TOHP	Trials of Hypertension Prevention
TRANSCEND	Telmisartan Randomized AssessmeNt Study in ACE INtolerant subjects with cardiovascular Disease
UK	urinary potassium
UL	Tolerable Upper Intake Level
UNa	urinary sodium
USDA	U.S. Department of Agriculture
WHO	World Health Organization

Appendix B

Committee Member
Biographical Sketches

Brian L. Strom, M.D., M.P.H. (*Chair*), is the Executive Vice Dean for Institutional Affairs in the Perelman School of Medicine at the University of Pennsylvania. He also is George S. Pepper Professor of Public Health and Preventive Medicine, professor of biostatistics and epidemiology, professor of medicine, and professor of pharmacology. Dr. Strom previously served as president of the Association of Clinical Research Training and currently is principal investigator (PI) or co-PI for eight National Institutes of Health (NIH)-funded clinical research training programs. Dr. Strom's interests span many areas of clinical epidemiology, but his major research interest is the field of pharmacoepidemiology, i.e., the application of epidemiologic methods to the study of drug use and effects. He is best known as a founder of the field of pharmacoepidemiology, and a pioneer in using large automated databases for research. He is editor of the field's major text, and is editor-in-chief for *Pharmacoepidemiology and Drug Safety*, the official journal of the International Society for Pharmacoepidemiology. Dr. Strom is a former member of the Board of Regents of the American College of Physicians, and the Boards of Directors of the American Society for Clinical Pharmacology & Therapeutics, the American College of Epidemiology, and the Association for Patient-Oriented Research. Dr. Strom earned his M.D. from the Johns Hopkins University School of Medicine and M.P.H. in epidemiology at the University of California, Berkeley. Dr. Strom has been a member of the Institute of Medicine since 2001.

Cheryl A. M. Anderson, Ph.D., M.P.H., is associate professor in the Department of Family and Preventive Medicine at the University of California,

San Diego. Before this appointment she was an assistant professor in the Department of Epidemiology at the Johns Hopkins Bloomberg School of Public Health. Dr. Anderson's research centers on nutrition-related issues in chronic disease prevention in minority and underserved populations. She is the PI of the National Heart, Lung, and Blood Institute (NHLBI)-funded study of the effects of dietary sodium and potassium intake on subclinical and clinical cardiovascular disease. She is a co-investigator on the National Institute of Diabetes and Digestive and Kidney Diseases–funded national, multicenter Chronic Renal Insufficiency Cohort Study, which aims to identify risk factors and mechanisms of progressive renal disease and cardiovascular events in individuals with chronic kidney disease. She also is a co-investigator on the NHLBI-funded OMNI-Carb study, a randomized feeding trial that compares the effects of type (glycemic index) and amount of carbohydrate on cardiovascular risk factors. Dr. Anderson is PI of a study testing a unique biomarker (using carbon isotopic data) of intake of sweets (funded by an Innovation Grant Award from the Johns Hopkins Bloomberg School of Public Health). She is a member of the Food and Nutrition Board and has served on the IOM Committee on Dietary Supplement Use in the Military and the Committee on Strategies to Reduce Sodium Intake. She has an M.P.H. from the University of North Carolina at Chapel Hill and a Ph.D. in nutritional sciences from the University of Washington School of Public Health and Community Medicine.

Jamy Ard, M.D., is associate professor in the Department of Epidemiology and Prevention and the Department of Medicine at Wake Forest University Baptist Medical Center. He also is co-director of the Wake Forest Baptist Health Weight Management Center, directing medical weight management programs. Dr. Ard received an M.D. and completed internal medicine residency training at Duke University Medical Center. He also received formal training in clinical research as a fellow at the Center for Health Services Research in Primary Care at the Durham VA Medical Center. Dr. Ard has more than 15 years of experience in clinical nutrition and obesity. Before joining the faculty at Wake Forest in 2012, Dr. Ard spent 9 years at the University of Alabama at Birmingham, where he served as medical director of UAB's EatRight Weight Management Services, vice-chair for clinical care in the Department of Nutrition Sciences, and associate dean for clinical affairs in the School of Health Professions. Dr. Ard's research interests include clinical management of obesity and strategies to improve cardiometabolic risk using lifestyle modification. He has a special interest in the African American population and in developing strategies for behavior modification that are culturally appropriate for this group. He has been conducting research on lifestyle modification since 1995 and has worked on several NIH-funded multicenter trials, including Dietary Approaches to Stop Hypertension (DASH), DASH-Sodium, and Weight Loss Maintenance Trial. His work has been published in numerous scientific journals and he has been a featured presenter at several conferences and workshops dealing with obesity. Currently, he also is serving as a member of the Expert Panel on the Identification, Evaluation, and Treatment of Overweight and Obesity in Adults. This group, sponsored by NHLBI, is revising the 1998 guidelines for the clinical management of overweight and obesity.

Kirsten Bibbins-Domingo, Ph.D., M.D., M.A.S., is Associate Professor of Medicine and of Epidemiology and Biostatistics at the University of California, San Francisco (UCSF). She also is the Director of the UCSF Center for Vulnerable Populations at San Francisco General Hospital. A general internist at San Francisco General Hospital and faculty member in the Division of General Internal Medicine, Dr. Bibbins-Domingo is a cardiovascular epidemiologist who has published extensively on the development of heart disease in young adults and on race/ethnic and income differences in manifestations of heart disease. Her current work focuses on understanding the interaction between social, behavioral, and biological factors that place vulnerable groups at risk of cardiovascular disease early in life, and population-wide policy-level interventions that may prevent disease in these groups. She is an inducted member of the American Society for Clinical Investigation and a member of the U.S. Preventive Services Task Force. She received her undergraduate degree from Princeton University, an M.D. from UCSF School of Medicine, and a Ph.D. in biochemistry from UCSF.

Nancy R. Cook, Sc.D., is professor in the Department of Medicine at Harvard Medical School and the Brigham & Women's Hospital and Harvard Medical School, and Professor of Epidemiology at the Harvard School of Public Health. Dr. Cook is a biostatistician involved in the design, conduct, and analysis of several large randomized trials, including the Women's Health Study, the Physicians' Health Study, the Women's Anti-Oxidant Cardiovascular Study, and the VITamin D and OmegA-3 TriaL (VITAL). She leads the Trials of Hypertension Prevention (TOHP) Follow-up Study, an observational follow-up of participants in Phases I and II of TOHP. This study focuses on the long-term effects of weight loss and sodium reduction interventions on subsequent cardiovascular disease, as well as on the observational effects of average sodium intake and intentional weight loss. Dr. Cook also works in the field of predictive modeling of observational data, for detecting gene–gene and gene–environment interactions, as well as for developing risk prediction scores using clinical biomarkers. She received an Sc.D. from the Harvard School of Public Health.

Mary Kay Fox, M.Ed., is Senior Fellow and Area Leader for nutrition policy research at Mathematica Policy Research, Inc. Ms. Fox has more than 25 years of research experience with expertise in dietary intake assessment and the evaluation of nutrition programs and policies to promote health and prevent disease. Ms. Fox has used data from the National Health and Nutrition Examination Survey and other national data sets to assess sodium intakes across the life span and among low- and high-income population groups. She also has assessed the sodium content of meals offered and served in federally funded school meal programs and how these meals contribute to children's dietary intakes and obesity risk. Ms. Fox is currently directing an evaluation of the Harlem Children's Zone Healthy Living Initiative, an obesity prevention and treatment initiative that is being implemented in early child care programs, charter schools, and after-school programs. She also is leading the FNS WIC-Medicaid Study-II, which will assess the impact of WIC participation on birth outcomes and Medicaid costs. Ms. Fox previously served on the IOM Committee to Review Child and Adult Care Food Program Meal Requirements, as well as the Committee on Nutrition Standards for the National School Lunch and Breakfast Programs. She holds an M.Ed. in nutrition from Tufts University.

Niels Graudal, M.D., Dr.M.Sc., is an attending physician in the Department of Rheumatology at the Copenhagen University Hospital in Rigshospitalet, Denmark. Before this appointment he was a registrar and senior registrar at university departments in the Copenhagen area, including internal medicine, gastroenterology, cardiology, pulmonary medicine, allergic diseases, hepatology, hematology, nephrology, infectious diseases, and rheumatology. Dr. Graudal has conducted and published periodic meta-analyses and reviews of the data on the health effects of dietary sodium intake from clinical trials and epidemiological studies. In addition, his recent research centers on sarcoidosis and the treatment of rheumatoid arthritis. He has an M.D. from the University of Aarhus in Denmark.

Jiang He, M.D., Ph.D., is Joseph S. Copes Chair and Professor in the Department of Epidemiology at Tulane University. His research interests include the study of the etiology and prevention of cardiovascular disease, kidney disease, and stroke; gene and environment interaction on hypertension and other cardiovascular disease; global health; gender and ethnic disparities in health; and translational and implementation research. He has published extensively on a variety of topics, including cardiovascular disease risk factors. Dr. He previously served on the IOM Committee on Public Health Priorities to Reduce and Control Hypertension and the IOM Committee on Secondhand Smoke Exposure and Acute Coronary Events. Dr. He earned a Ph.D. from the Johns Hopkins Bloomberg School of Public

Health, a Dr.Med.Sc. from Peking Union Medical College, and an M.D. from Jiangxi Medical College in Jiangxi, China.

Joachim Ix, M.D., M.A.S., F.A.S.N., is Associate Professor of Medicine and a nephrologist and epidemiologist at the University of California, San Diego, School of Medicine. His research applies epidemiology and biostatistics to the understanding of mechanisms by which alterations in glucose and mineral metabolism contribute to cardiovascular disease risk among persons with kidney disease. His recent work has focused on the consequences of altered concentrations of fetuin-A, a hepatic secretory protein that simultaneously inhibits vascular calcification and promotes insulin resistance. In addition, he studies cystatin-C, a novel endogenous measure of kidney function, to determine whether this measure might provide new insights into cardiovascular disease mechanisms among persons with early kidney function decline. Dr. Ix received an M.D. from the University of Chicago Pritzker School of Medicine and an M.A.S. in epidemiology and biostatistics at UCSF.

Stephen E. Kimmel, M.D., M.S.C.E., is Professor of Medicine and Epidemiology at the University of Pennsylvania's Perelman School of Medicine. Dr. Kimmel's research focuses on cardiac pharmacoepidemiology, with a particular focus on the effects and proper use of drugs and devices for patients with cardiac disease. He is currently PI on a number of grants evaluating cardiac medications, including a study examining the effects of genetic polymorphisms on anticoagulation control. He also is the PI of a randomized trial designed to test economic and behavioral interventions to improve adherence with warfarin treatment and the PI of a multicenter clinical trial examining the effects of genetic-based dosing of warfarin. He has an M.D. from the New York University School of Medicine and a master of science in clinical epidemiology from the University of Pennsylvania.

Alice H. Lichtenstein, D.Sc., is Stanley N. Gershoff Professor of Nutrition Science and Policy in the Friedman School at Tufts University and Director and Senior Scientist of the Cardiovascular Nutrition Laboratory at the U.S. Department of Agriculture's Jean Mayer Human Nutrition Research Center on Aging, also at Tufts University. She holds secondary appointments as an associate faculty member in the Institute for Clinical Research and Health Policy Studies at Tufts Medical Center and as Professor of Medicine at Tufts University School of Medicine. Dr. Lichtenstein's research group focuses on assessing the interplay between diet and heart disease risk factors. Past and current work includes addressing in postmenopausal females and older males, issues related to *trans* fatty acids, soy protein and isoflavones, sterol/stanolesters, and novel vegetable oils differing in fatty acid profile and glycemic index. Selected issues are investigated in animal models and cell systems with the aim of determining the mechanisms by which dietary factors alter cardiovascular disease risk. Additional work is focused on population-based studies to address the relationship of cholesterol homeostasis and nutrient biomarkers on cardiovascular disease risk, and on the application of systematic review methods to the field of nutrition. She served on the IOM Dietary Reference Intake macronutrient panel, the IOM Food Forum and the IOM Committee on Front-of-Package Nutrition Rating Systems and Symbols: Promoting Healthier Choices. Dr. Lichtenstein completed her undergraduate work at Cornell University, holds a master's degree from the Pennsylvania State University, and master's and doctoral degrees from Harvard University.

Myron Weinberger, M.D., is Emeritus Professor of Medicine and Director of the Hypertension Research Center at the Indiana University School of Medicine and Editor-in-Chief of the *Journal of the American Society of Hypertension.* Dr. Weinberger's specialty is internal medicine and hypertension. His research interests include the relationship between salt intake and high blood pressure. In addition to serving on numerous scientific review committees, Dr. Weinberger is a member of several editorial boards, including those of *Hypertension* and the *Journal of the American College of Nutrition.* He has published more than 200 scientific articles on hypertension, many of which relate to the roles of sodium and/or potassium. Dr. Weinberger received the Robert Tigerstedt Award from the American Society of Hypertension and the Page-Bradley Lifetime Achievement Award from the Council for High Blood Pressure Research of the American Heart Association for his research in hypertension. He received an M.D. from the Indiana University School of Medicine.

Appendix C

Open Session Agendas

Tuesday December 4, 2012

OPEN SESSION

3:00 p.m.   **Presentations by CDC Representatives**
Robert K. Merritt
Mary Cogswell
Janelle Gunn

**A WORKSHOP ON PERSPECTIVES ON
DIETARY SODIUM AND HEALTH**
December 5, 2012

National Academy of Sciences
2101 Constitution Avenue, NW
Washington, DC 20418
Room 125

Wednesday, December 5, 2012

8:00 a.m.   **Registration**

8:25   **Welcome**
Brian Strom, *Committee Chair*

Session 1 – Government Efforts Concerning Sodium Intake

8:30   Session 1 Introduction
            Moderator: BRIAN STROM

8:35   Current CDC Efforts Concerning Sodium Intake
            ROBERT K. MERRITT, *Centers for Disease Control and
            Prevention*

            Cardiovascular Risk Reduction in Adults: The Lifestyle
            Workgroup Background and Methods
            JANET DE JESUS, *National Heart, Lung, and Blood Institute*

            FDA's Activities in Sodium Reduction
            JEREMIAH FASANO, *Food and Drug Administration*

9:10   Questions for Session 1 Speakers

Session 2 – Setting the Stage for Examining Sodium Intake in Populations

9:20   Session 2 Introduction
            Moderator: JOE IX

9:25   Salt Sensitivity: Mechanisms, Diagnosis, and Clinical
            Relevance
            MATT WEIR, *University of Maryland*

9:50   Methodological Considerations to Assess Dietary Sodium
            Intake in the Population Using What We Eat in America,
            National Health and Nutrition Examination Survey
            (NHANES)
            ALANA MOSHFEGH, *USDA Agricultural Research Service*

10:25   Sodium Reduction Initiatives in the Americas
            BRANKA LEGETIC, *Pan American Health Organization*

10:50   Questions for Session 2 Speakers

11:10   Break

Session 3 – Approaches to Reviewing Evidence

11:25   Session 3 Introduction
         Moderator: ALICE LICHTENSTEIN

11:30   Lesson Learned on Conducting Nutrition Systematic Reviews
         JOSEPH LAU, *Brown University*

11:50   Questions for Session 3 Speakers

12:00 p.m. Lunch

Session 4 – Health Effects Associated with Lowering Sodium Intake in the Population

1:00   Session 4 Introduction
         Moderator: KIRSTEN BIBBINS-DOMINGO

1:05   Low Versus Moderate Sodium Intake to Reduce
         Cardiovascular Events
         MARTIN O'DONNELL, *McMaster University, Canada National
         University of Ireland, Dublin*

1:30   Dietary Sodium and Cardiovascular Outcomes: The Evidence
         MICHAEL H. ALDERMAN, *Albert Einstein College of Medicine*

1:55   A Review of Heath Benefits of Lowering Sodium Intake in
         the United States
         LAWRENCE APPEL, *Johns Hopkins Medical Institutions*

2:20   Importance of Mineral Interactions in Heart and Bone
         Health: Sodium, Potassium, Calcium, and Magnesium
         CONNIE WEAVER, *Purdue University*

2:45   The Pleiotropic Effects of Dietary Sodium
         MERLIN THOMAS, *Baker IDI Heart and Diabetes Institute*

3:10   Balancing the Evidence Regarding Sodium and Health
         SALIM YUSUF, *McMaster University, Canada*

3:35   **Sodium Intake's Physiological Range: Perturbation's
            Pathological Consequences**
            DAVID MCCARRON, *Department of Nutrition, University of
            California, Davis*

4:00   **Questions for Session 4 Speakers**

            Session 5 – Public Comments and Review of the Day

4:30   **Public Comments (5 minutes each)**
            Moderator: BRIAN STROM

5:30   **Adjourn**

Appendix D

Biomarkers Figure

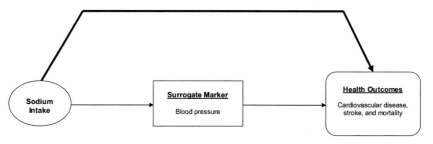

FIGURE D-1 Relationships between sodium intake, surrogate marker (blood pressure) and health outcomes. Surrogate markers are a laboratory measurement or physical sign used as an indicator of a health outcome. Health outcomes represent incidence (e.g., development) of a disease or mortality.
SOURCE: Adapted from Russell et al., 2009. Reprinted with permission from *American Journal of Clinical Nutrition* (2009; 89:728-733).

REFERENCE

Russell, R., M. Chung, E. M. Balk, S. Atkinson, E. L. Giovannucci, S. Ip, A. H. Lichtenstein, S. T. Mayne, G. Raman, A. Catharine Ross, T. A. Trikalinos, K. P. West, Jr., and J. Lau. 2009. Opportunities and challenges in conducting systematic reviews to support the development of nutrient reference values: Vitamin A as an example. *American Journal of Clinical Nutrition* 89(3):728-733.

Appendix E

Literature Search Strategy

A literature search strategy was conducted to select the scientific literature published after 2003 for the committee's review and answer the question in the statement of task related to the potential associations of sodium intake and health outcomes. The online databases usedfor these search were Cochrane Database of Systematic Reviews, Embase, MedLine, PubMed, and Web of Science. A broad search to include all health outcomes and a number of searches targeted at specific outcomes identified by the committee were conducted. The specific outcomes were cardiovascular disease, heart failure, hypertension, myocardial infarction, diabetes, mortality, stroke, bone disease, fractures, falls, myocardial infarction, headaches, kidney stones, chronic kidney disease, skin reactions, immune function, thyroid disease, and cancer. Table E-1 presents the search conducted in the MedLine database. The searches were conducted in consultation with the staff librarians at the George E. Brown Jr. Library of the National Academies.

TABLE E-1 Example (MedLine) of Searches to Identify Relevant Literature on Sodium Intake and Health Outcomes

Search No.	Search Terms	Number of Hits
1	Sodium, Dietary/or Sodium Chloride, Dietary/or Diet, Sodium-Restricted/	4,022
2	1 and (health or disease$ or condition$).mp. [mp=title, abstract, original title, name of substance word, subject heading word, protocol supplementary concept, rare disease supplementary concept, unique identifier]	1,936
3	1 and hypertension/	1,445
4	1 and (cardiovascular diseases/or coronary disease/or death, sudden/)	289
5	1 and heart failure/	130
6	1 and mortality/	10
7	1 and stroke/	80
8	1 and (fractures, bones/or accidental falls/)	1
9	1 and myocardial infarction/	22
10	1 and headache/	3
11	1 and kidney calculi/	17
12	1 and skin manifestations/	0
13	1 and thyroid diseases/	28
14	1 and immunity/	0
15	1 and diabetes mellitus/	21
16	1 and (kidney failure, chronic/or kidney diseases/)	242
17	1 and neoplasms/	13
	Total	2,687
	Total without animal or in vitro studies	1,938

Appendix F

Presentation of Results—Evidence Tables

This appendix contains tables summarizing the studies reviewed by the committee organized by health outcome, type of study, and alphabetically by first author. Studies that reviewed multiple health outcomes are listed in more than one table. Some studies included in these tables were not included in the committee's final review and assessment of the evidence for associations between dietary sodium intake and outcomes discussed in Chapter 4 due to inability to meet the inclusion criteria. As described in Chapter 2, the committee reviewed peer-reviewed original research studies (excluding case studies and case series), published in the English language and published between January 1, 2003, and December 18, 2012, including those conducted in all countries and with all sample sizes, populations, and follow-up periods. Studies were excluded if they included only intermediate outcomes; did not use a food frequency questionnaire, 24-hour recall, dietary diary, or urine analysis methods to estimate dietary sodium intake; did not calculate numerical sodium levels; or did not analyze the independent association between sodium and a health outcome. Studies that reviewed multiple health outcomes are included in more than one table.

LIST OF TABLES

REFERENCES

Arakawa, K., Y. Matsushita, H. Matsuo, N. Ikeda, M. Iwashita, and K. Kuramoto. 2009. Examination of the efficiency of salt taste preference questionnaire in hypertensive patients—Results from post marketing surveillance of Olmetec (R) tablets and Calblock (R) tablets. *Rinsho Iyaku* 25:723-734.

Arcand, J., J. Ivanov, A. Sasson, V. Floras, A. Al-Hesayen, E. R. Azevedo, S. Mak, J. P. Allard, and G. E. Newton. 2011. A high-sodium diet is associated with acute decompensated heart failure in ambulatory heart failure patients: A prospective follow-up study. *American Journal of Clinical Nutrition* 93(2):332-337.

Baune, B. T., Y. Aljeesh, and R. Bender. 2005. Factors of non-compliance with the therapeutic regimen among hypertensive men and women: A case-control study to investigate risk factors of stroke. *European Journal of Epidemiology* 20(5):411-419.

Chang, H. Y., Y. W. Hu, C. S. Yue, Y. W. Wen, W. T. Yeh, L. S. Hsu, S. Y. Tsai, and W. H. Pan. 2006. Effect of potassium-enriched salt on cardiovascular mortality and medical expenses of elderly men. *American Journal of Clinical Nutrition* 83(6):1289-1296.

Cohen, H. W., S. M. Hailpern, J. Fang, and M. H. Alderman. 2006. Sodium intake and mortality in the NHANES II follow-up study. *American Journal of Medicine* 119(3):275. e7-275.e14.

Cohen, H. W., S. M. Hailpern, and M. H. Alderman. 2008. Sodium intake and mortality follow-up in the Third National Health and Nutrition Examination Survey (NHANES III). *Journal of General Internal Medicine* 23(9):1297-1302.

Cook, N. R., J. A. Cutler, E. Obarzanek, J. E. Buring, K. M. Rexrode, S. K. Kumanyika, L. J. Appel, and P. K. Whelton. 2007. Long term effects of dietary sodium reduction on cardiovascular disease outcomes: Observational follow-up of the trials of hypertension prevention (TOHP). *British Medical Journal* 334(7599):885-888.

Cook, N. R., E. Obarzanek, J. A. Cutler, J. E. Buring, K. M. Rexrode, S. K. Kumanyika, L. J. Appel, and P. K. Whelton. 2009. Joint effects of sodium and potassium intake on subsequent cardiovascular disease: The trials of hypertension prevention follow-up study. *Archives of Internal Medicine* 169(1):32-40.

Costa, A. P. R., R. C. S. de Paula, G. F. Carvalho, J. P. Araújo, J. M. Andrade, O. L. R. de Almeida, E. C. de Faria, W. M. Freitas, O. R. Coelho, J. A. F. Ramires, J. C. Quinaglia e Silva, and A. C. Sposito. 2012. High sodium intake adversely affects oxidative-inflammatory response, cardiac remodelling and mortality after myocardial infarction. *Atherosclerosis* 222(1):284-291.

Daimon, M., H. Sato, S. Sasaki, S. Toriyama, M. Emi, M. Muramatsu, S. C. Hunt, P. N. Hopkins, S. Karasawa, K. Wada, Y. Jimbu, W. Kameda, S. Susa, T. Oizumi, A. Fukao, I. Kubota, S. Kawata, and T. Kato. 2008. Salt consumption-dependent association of the GNB3 gene polymorphism with type 2 DM. *Biochemical & Biophysical Research Communications* 374(3):576-580.

Dong, J., Y. Li, Z. Yang, and J. Luo. 2010. Low dietary sodium intake increases the death risk in peritoneal dialysis. *Clinical Journal of the American Society of Nephrology* 5(2):240-247.

Ekinci, E. I., S. Clarke, M. C. Thomas, J. L. Moran, K. Cheong, R. J. MacIsaac, and G. Jerums. 2011. Dietary salt intake and mortality in patients with type 2 diabetes. *Diabetes Care* 34(3):703-709.

Gardener, H., T. Rundek, C. B. Wright, M. S. V. Elkind, and R. L. Sacco. 2012. Dietary sodium and risk of stroke in the Northern Manhattan Study. *Stroke* 43(5):1200-1205.

Geleijnse, J. M., J. C. M. Witteman, T. Stijnen, M. W. Kloos, A. Hofman, and D. E. Grobbee. 2007. Sodium and potassium intake and risk of cardiovascular events and all-cause mortality: The Rotterdam Study. *European Journal of Epidemiology* 22(11):763-770.

Heerspink, H. J. L., F. A. Holtkamp, H. H. Parving, G. J. Navis, J. B. Lewis, E. Ritz, P. A. De Graeff, and D. De Zeeuw. 2012. Moderation of dietary sodium potentiates the renal and cardiovascular protective effects of angiotensin receptor blockers. *Kidney International* 82(3):330-337.

Hu, G., P. Jousilahti, M. Peltonen, J. Lindstrom, and J. Tuomilehto. 2005. Urinary sodium and potassium excretion and the risk of type 2 diabetes: A prospective study in Finland. *Diabetologia* 48(8):1477-1483.

Jafar, T. H. 2006. Blood pressure, diabetes, and increased dietary salt associated with stroke— results from a community-based study in Pakistan. *Journal of Human Hypertension* 20(1):83-85.

Kono, Y., S. Yamada, K. Kamisaka, A. Araki, Y. Fujioka, K. Yasui, Y. Hasegawa, and Y. Koike. 2011. Recurrence risk after noncardioembolic mild ischemic stroke in a Japanese population. *Cerebrovascular Diseases* 31(4):365-372.

Larsson, S. C., M. J. Virtanen, M. Mars, S. Männistö, P. Pietinen, D. Albanes, and J. Virtamo. 2008. Magnesium, calcium, potassium, and sodium intakes and risk of stroke in male smokers. *Archives of Internal Medicine* 168(5):459-465.

Lazarević, K., A. Nagorni, D. Bogdanović, N. Rančić, L. Stošić, and S. Milutinović. 2011. Dietary micronutrients and gastric cancer: Hospital based study. *Central European Journal of Medicine* 6(6):783-787.

Lee, S. A., D. Kang, K. N. Shim, J. W. Choe, W. S. Hong, and H. Choi. 2003. Effect of diet and *Helicobacter pylori* infection to the risk of early gastric cancer. *Journal of Epidemiology* 13(3):162-168.

Lennie, T. A., E. K. Song, J. R. Wu, M. L. Chung, S. B. Dunbar, S. J. Pressler, and D. K. Moser. 2011. Three gram sodium intake is associated with longer event-free survival only in patients with advanced heart failure. *Journal of Cardiac Failure* 17(4):325-330.

McCausland, F. R., S. S. Waikar, and S. M. Brunelli. 2012. Increased dietary sodium is independently associated with greater mortality among prevalent hemodialysis patients. *Kidney International* 82(2):204-211.

Murata, A., Y. Fujino, T. M. Pham, T. Kubo, T. Mizoue, N. Tokui, S. Matsuda, and T. Yoshimura. 2010. Prospective cohort study evaluating the relationship between salted food intake and gastrointestinal tract cancer mortality in Japan. *Asia Pacific Journal of Clinical Nutrition* 19(4):564-571.

Nagata, C., N. Takatsuka, N. Shimizu, and H. Shimizu. 2004. Sodium intake and risk of death from stroke in Japanese men and women. *Stroke* 35(7):1543-1547.

O'Donnell, M. J., S. Yusuf, A. Mente, P. Gao, J. F. Mann, K. Teo, M. McQueen, P. Sleight, A. M. Sharma, A. Dans, J. Probstfield, and R. E. Schmieder. 2011. Urinary sodium and potassium excretion and risk of cardiovascular events. *Journal of the American Medical Association* 306(20):2229-2238.

Parrinello, G., P. Di Pasquale, G. Licata, D. Torres, M. Giammanco, S. Fasullo, M. Mezzero, and S. Paterna. 2009. Long-term effects of dietary sodium intake on cytokines and neurohormonal activation in patients with recently compensated congestive heart failure. *Journal of Cardiac Failure* 15(10):864-873.

Paterna, S., P. Gaspare, S. Fasullo, F. M. Sarullo, and P. Di Pasquale. 2008. Normal-sodium diet compared with low-sodium diet in compensated congestive heart failure: Is sodium an old enemy or a new friend? *Clinical Science* 114(3):221-230.

Paterna, S., G. Parrinello, S. Cannizzaro, S. Fasullo, D. Torres, F. M. Sarullo, and P. Di Pasquale. 2009. Medium term effects of different dosage of diuretic, sodium, and fluid administration on neurohormonal and clinical outcome in patients with recently compensated heart failure. *American Journal of Cardiology* 103(1):93-102.

Paterna, S., S. Fasullo, G. Parrinello, S. Cannizzaro, I. Basile, G. Vitrano, G. Terrazzino, G. Maringhini, F. Ganci, S. Scalzo, F. M. Sarullo, G. Cice, and P. Di Pasquale. 2011. Short-term effects of hypertonic saline solution in acute heart failure and long-term effects of a moderate sodium restriction in patients with compensated heart failure with New York Heart Association class III (class C) (SMAC-HF study). *American Journal of the Medical Sciences* 342(1):27-37.

Peleteiro, B., C. Lopes, C. Figueiredo, and N. Lunet. 2011. Salt intake and gastric cancer risk according to *Helicobacter pylori* infection, smoking, tumour site and histological type. *British Journal of Cancer* 104(1):198-207.

Pelucchi, C., I. Tramacere, P. Bertuccio, A. Tavani, E. Negri, and C. La Vecchia. 2009. Dietary intake of selected micronutrients and gastric cancer risk: An Italian case-control study. *Annals of Oncology* 20(1):160-165.

Rodrigues, S. L., M. P. Baldo, R. de Sa Cunha, R. V. Andreao, M. Del Carmen Bisi Molina, C. P. Goncalves, E. M. Dantas, and J. G. Mill. 2009. Salt excretion in normotensive individuals with metabolic syndrome: A population-based study. *Hypertension Research—Clinical & Experimental* 32(10):906-910.

Roy, M. S., and M. N. Janal. 2010. High caloric and sodium intakes as risk factors for progression of retinopathy in type 1 diabetes mellitus. *Archives of Ophthalmology* 128(1):33-39.

Shikata, K., Y. Kiyohara, M. Kubo, K. Yonemoto, T. Ninomiya, T. Shirota, Y. Tanizaki, Y. Doi, K. Tanaka, Y. Oishi, T. Matsumoto, and M. Iida. 2006. A prospective study of dietary salt intake and gastric cancer incidence in a defined Japanese population: The Hisayama study. *International Journal of Cancer* 119(1):196-201.

Sjödahl, K., C. Jia, L. Vatten, T. Nilsen, K. Hveem, and J. Lagergren. 2008. Salt and gastric adenocarcinoma: A population-based cohort study in Norway. *Cancer Epidemiology Biomarkers and Prevention* 17(8):1997-2001.

Stolarz-Skrzypek, K., T. Kuznetsova, L. Thijs, V. Tikhonoff, J. Seidlerová, T. Richart, Y. Jin, A. Olszanecka, S. Malyutina, E. Casiglia, J. Filipovský, K. Kawecka-Jaszcz, Y. Nikitin, and J. A. Staessen. 2011. Fatal and nonfatal outcomes, incidence of hypertension, and blood pressure changes in relation to urinary sodium excretion. *Journal of the American Medical Association* 305(17):1777-1785.

Strumylaite, L., J. Zickute, J. Dudzevicius, and L. Dregval. 2006. Salt-preserved foods and risk of gastric cancer. *Medicina* 42(2):164-170.

Takachi, R., M. Inoue, T. Shimazu, S. Sasazuki, J. Ishihara, N. Sawada, T. Yamaji, M. Iwasaki, H. Iso, Y. Tsubono, and S. Tsugane. 2010. Consumption of sodium and salted foods in relation to cancer and cardiovascular disease: The Japan Public Health Center-based prospective study. *American Journal of Clinical Nutrition* 91(2):456-464.

Teramoto, T., R. Kawamori, S. Miyazaki, and S. Teramukai. 2011. Sodium intake in men and potassium intake in women determine the prevalence of metabolic syndrome in Japanese hypertensive patients: OMEGA Study. *Hypertension Research* 34(8):957-962.

Thomas, M. C., J. Moran, C. Forsblom, V. Harjutsalo, L. Thorn, A. Ahola, J. Wadén, N. Tolonen, M. Saraheimo, D. Gordin, and P. H. Groop. 2011. The association between dietary sodium intake, ESRD, and all-cause mortality in patients with type 1 diabetes. *Diabetes Care* 34(4):861-866.

Tikellis, C., R. J. Pickering, D. Tsorotes, V. Harjutsalo, L. Thorn, A. Ahola, J. Waden, N. Tolonen, M. Saraheimo, D. Gordin, C. Forsblom, P. H. Groop, M. E. Cooper, J. Moran, and M. C. Thomas. 2013. Association of dietary sodium intake with atherogenesis in experimental diabetes and with cardiovascular disease in patients with type 1 diabetes. *Clinical Science* 124(10):617-626.

Tsugane, S., S. Sasazuki, M. Kobayashi, and S. Sasaki. 2004. Salt and salted food intake and subsequent risk of gastric cancer among middle-aged Japanese men and women. *British Journal of Cancer* 90(1):128-134.

Umesawa, M., H. Iso, C. Date, A. Yamamoto, H. Toyoshima, Y. Watanabe, S. Kikuchi, A. Koizumi, T. Kondo, Y. Inaba, N. Tanabe, and A. Tamakoshi. 2008. Relations between dietary sodium and potassium intakes and mortality from cardiovascular disease: The Japan Collaborative Cohort study for evaluation of cancer risks. *American Journal of Clinical Nutrition* 88(1):195-202.

van den Brandt, P. A., A. A. M. Botterweck, and R. A. Goldbohm. 2003. Salt intake, cured meat consumption, refrigerator use and stomach cancer incidence: A prospective cohort study (Netherlands). *Cancer Causes and Control* 14(5):427-438.

Yang, Q., T. Liu, E. V. Kuklina, W. D. Flanders, Y. Hong, C. Gillespie, M. H. Chang, M. Gwinn, N. Dowling, M. J. Khoury, and F. B. Hu. 2011. Sodium and potassium intake and mortality among US adults: Prospective data from the Third National Health and Nutrition Examination Survey. *Archives of Internal Medicine* 171(13):1183-1191.

Zhang, Z., and X. Zhang. 2011. Salt taste preference, sodium intake and gastric cancer in China. *Asian Pacific Journal of Cancer Prevention* 12(5):1207-1210.

TABLE F-1 Evidence Tables: CVD/Stroke/Mortality Randomized
Controlled Trials

Citation	Population Studied	Study Design	Intervention/ Control	Sample Size
Chang et al., 2006	Five kitchens at a veterans retirement home, >40 y	RCT	*Intervention* K-enriched salt (49 percent sodium chloride, 49 percent potassium chloride)	*Intervention* 768 (mean age: 74.8±7.1 y)
				Control 1,213 (mean age: 74.9±6.7 y)
			Control Regular salt	

NOTES FOR TABLES F-1 THROUGH F-10

*Indicates significance.

Sodium intake presented as mmol was converted to mg using 23 mg/mmol.
ACE, angiotensin-converting enzyme; ACM, all-cause mortality; ADHF,
acute decompensated heart failure; amt, amount; ARB, angiotension receptor blockers; ARR, absolute risk reduction; BMI, body mass index; BP,
blood pressure; Ca, calcium; CHD, coronary heart disease; CHF, congestive
heart failure; CI, confidence interval; CKD, chronic kidney disease; CVD,
cardiovascular disease; d, day; dl, deciliter; DM, diabetes mellitus; ESRD,
end-stage renal disease; FFQ, food frequency questionnaire; g, grams; h,
hour; HDL, high-density-lipoprotein cholesterol; HR, hazard ratio; HSS,
hypertonic saline solution; IDNT, Irbesartan Diabetic Nephropathy Trial;
IHD, ischemic heart disease; IS, ischemic stroke; K, potassium; Kt/V, measurement of urea removal; L, liter; LDL, low-density-lipoprotein cholesterol;
LVEF, left ventricular ejection fraction; mg, milligrams; MI, myocardial

Sodium Exposure (method and level)	Co-intervention	Blinding	Follow-up Period	Health Outcome	Results
Calculated from number of meals served and amount of salt used per day Intervention 3,800 mg *Control* 5,200 mg Urine electrolyte data available for about 25% of the subjects	N/A	Single blind (participants)	2.6 y (length of intervention and mean follow-up)	ACM CVD mortality	Use of K-enriched salt associated with significant reduction in CVD mortality; however, may be primarily due to increased K intake *Intervention vs. Control* ACM HR=0.90, CI: 0.79, 1.06 *CVD mortality* *Intervention: HR=0.59, CI: 0.37, 0.95

infarction; ml, milliliter; mm HG, millimeters mercury; mo, month; Na, sodium; N/A, not applicable; NCI, National Cancer Institute; NHANES, National Health and Nutrition Examination Survey; NS, not significant; NYHA, New York Heart Association; ONTARGET, ONgoing Telmisartan Alone and in combination with Ramipril Global Endpoint Trial; OR, odds ratio; Q, quartile/quintile; RAAS, rennin-angiotensin-aldosterone system; RCT, randomized controlled trial; RENAAL, Reduction of Endpoints in NIDDM with the Angiotensiin II Antagonist Losartan Study; RR, relative risk; sat. fat, saturated fat; SD, standard deviation; T, tertile; TOHP, Trials of Hypertension Prevention; TRANSCEND, Telmisartan Randomized AssessmeNt Study in ACE iNtolerant subjects with cardiovascular Disease; UK, urinary potassium; UNa, urinary sodium; USDA, U.S. Department of Agriculture; vs., versus; wk, week; y, year.

TABLE F-2 Evidence Tables: CVD/Stroke/Mortality Cohort Studies

Citation	Population Studied	Study Design	Sample Size	Sodium Exposure (method and level)
Cohen et al., 2006	NHANES II, 30-74 y, without a history of CVD events	Prospective cohort	7,154	24-h dietary recall of Na intake at baseline Na intake levels ≥2,300 mg/d <2,300 mg/d
Cohen et al., 2008	NHANES III, ≥30 y, without a history of CVD events	Prospective cohort	8,699	24-h dietary recall of Na intake at baseline Na intake quartiles (intake level): Q1: <2,060 mg/d (1,501 mg/d) Q2: 2,060-2,921 mg/d (2,483 mg/d) Q3: 2,922-4,047 mg/d (3,441 mg/d) Q4: 4,048-9,946 mg/d (5,497 mg/d)
Cook et al., 2007	*Subset of 2,415 participants from 2 RCTs:* *TOHP (United States)* 30-54 y with diastolic BP 80-89 mmHg (prehypertensives) *TOHP II (United States)* 30-54 y with diastolic BP 83-89 mmHg and weighing 110-165% of their desirable weight (prehypertensives)	Prospective cohort	*TOHP I Intervention:* 327 (232 men; 95 women) *TOHP I Control:* 417 (299 men; 118 women) *TOHP II Intervention:* 1,191 (784 men; 407 women) *TOHP II Control:* 1,191 (782 men; 409 women)	*TOHP I* 24-h urine collection at baseline, 6, 12, and 18 mo. Net Na reduction at 18 mo.=1,012 mg/24 h (3,577 to 2,565 mg/d) *TOHP II* 24-h urine collection at baseline, 18, and 36 mo. Net Na reduction at 36 mo.=759 mg/24 h (4,225 to 3,466 mg/d)

Follow-up Period	Health Outcome	Confounders Adjusted for	Results
13.7 y	ACM CVD mortality CHD mortality Cerebrovascular disease mortality	Age, sex, race, smoking, alcohol use, systolic BP, antihypertensive treatment, BMI, education<high school, physical activity, dietary K, history of diabetes, serum cholesterol, calories	Lower Na intake associated with increased risk of ACM and CVD mortality For Na intake <2,300 mg/d: *ACM* *HR=1.28, CI: 1.10, 1.50, p=0.003 *CVD mortality* *HR=1.37, CI: 1.03, 1.81, p=0.03 *CHD mortality* HR=1.21, CI: 0.87, 1.68, p=0.25 *Cerebrovascular disease mortality* HR=1.78, CI: 0.89, 3.55, p=0.10
8.7±2.3 y	ACM CVD mortality	Age, sex, race, education, added table salt, exercise, alcohol use, current smoking, history of diabetes, history of cancer, systolic BP, cholesterol, dietary K, weight, treatment of hypertension, calories	Modest associations between lower Na intake and higher mortality *ACM* Q1: HR=1.24, CI: 0.91, 1.70 Q2: HR=1.30, CI: 0.96, 1.76 Q3: HR=1.06, CI: 0.81, 1.40 Q4: HR=1.00 p for Q1 vs. Q4=0.17 *CVD mortality* Q1: HR=1.80, CI: 1.05, 3.08 Q2: HR=1.94, CI: 1.32, 2.85 Q3: HR=1.48, CI: 0.82, 2.67 Q4: HR=1.00 *p for Q1 vs. Q4=0.03
10-15 y	Total mortality Incident CVD (MI, stroke, revascularization, or death due to CV cause)	Age, sex, race, trial, clinic, weight loss intervention	Lower Na excretion associated with reduced mortality and CVD *Mortality* *TOHP I* HR=0.81, CI: 0.52, 1.27, p=0.35 *CVD* *HR=0.75, CI: 0.57, 0.99, p=0.044

continued

TABLE F-2 Continued

Citation	Population Studied	Study Design	Sample Size	Sodium Exposure (method and level)
Cook et al., 2009	Subset of 2,275 participants from 2 RCTs: TOHP (United States) 30-54 y with diastolic BP 80-89 mmHg (prehypertensives) TOHP II (United States) 30-54 y with diastolic BP 83-89 mmHg and weighing 110-165% of their desirable weight (prehypertensives)	Prospective cohort	2,275	TOHP I 24-h urine collection at baseline, 6, 12, and 18 mo. TOHP II 24-h urine collection at baseline, 18, and 36 mo. Median excretion Overall: 3,634 mg/24 h (interquartile range: 2,921-4,462 mg/24 h) Men: 3,933 mg/24 h Women: 3,082 mg/24 h Na excretion quartiles (not provided)
Costa et al., 2012	Brasilia Heart Study subjects diagnosed with MI	Prospective cohort	372	62-item FFQ Na intake quantified using Brazilian Table of Food Composition, version 2 Na intake levels ≥1,200 mg/d=high; <1,200 mg/d=low
Dong et al., 2010	Chinese patients receiving peritoneal dialysis at single clinic, mean age=59.4±14.2 y	Retrospective cohort	305 (129 men; 176 women)	3-d dietary records completed by patients and checked by dietitian using food models Na calculated using computer software Na intake tertiles (average intake=1,820 mg/d; range=760-5,530 mg/d)

Follow-up Period	Health Outcome	Confounders Adjusted for	Results
10-15 y	CVD events	Age, sex, race/ethnicity, clinic, treatment assignment, education status, baseline weight, alcohol use, smoking, exercise, family history of CVD, changes in weight, smoking, exercise	Non-significant, relationship between UNa excretion and risk of CVD Q1: RR=1.0 Q2: RR=0.99, CI: 0.62, 1.58 Q3: RR=1.16, CI: 0.73, 1.84 Q4: RR=1.20, CI: 0.73, 1.97 p for trend=0.38 Adjusted for potassium excretion: *HR=1.42 (0.99-2.04) per 100 mmol/24 hr sodium, p=0.05
4 y	ACM	Age, sex, hypertension, diabetes, sedentarity, BMI	Risk of death was higher among individuals with Na intake >1,200 mg/d *Exp(B)=2.857, CI: 1.501, 5.437, p=0.01
31.4±13.7 mo.	ACM CVD mortality	Age, sex, BMI, history of DM or CVD, baseline total Kt/V, total creatinine clearance, mean arterial pressure, serum albumin, hemoglobin, Ca × phosphate, LDL	Lower Na intake associated with increased risk of ACM and CVD mortality *ACM* *HR=0.44, CI: 0.20, 0.95, p=0.04 *CVD mortality* *HR=0.11, CI: 0.03, 0.48, p=0.003

continued

TABLE F-2 Continued

Citation	Population Studied	Study Design	Sample Size	Sodium Exposure (method and level)
Ekinci et al., 2011	Patients attending a diabetes clinic in Melbourne, Australia 56% men; mean age= 64 y; median duration of diabetes=11 y; 47% obese	Prospective cohort	638	24-h urine collection Na excretion tertiles: T1: <3,450 mg/24 h T2: 3,450-4,784 mg/24 h T3: >4,784 mg/24 h
Gardener et al., 2012	Participants from the Northern Manhattan Study (New York), excluding those with stroke and MI (mean age=69±10 y, 64% women; 53% Hispanic; 24% African American; 21% white)	Population-based prospective cohort	2,657	Block National Cancer Institute FFQ Na intake calculated using DIETSYS software Na intake examined continuously (500 mg/d unit) and Na intake quartiles Q1: ≤1,500 mg/d Q2: 1,501-2,300 mg/d Q3: 2,301-3,999 mg/d Q4: 4,000-10,000 mg/d

Follow-up Period	Health Outcome	Confounders Adjusted for	Results
9.9 y (median)	ACM CVD mortality	Age, sex, duration of diabetes, atrial fibrillation, presence/severity of CKD	Lower Na excretion associated with increased risk of ACM and CVD mortality *ACM* *For every 2,300 mg/d rise in 24-h UNa excretion: HR=0.72, CI: 0.55, 0.94, p=0.017 *CVD mortality* *For every 2,300 mg/d rise in 24-h UNa excretion: sub-HR=0.65, CI: 0.44, 0.94, p=0.026
10 y (mean)	Vascular death Stroke incidence Stroke, MI, or vascular death MI	Age, sex, race/ethnicity, high school completion, diet, smoking, physical activity, alcohol consumption, daily energy, protein, fat, sat. fat, carbohydrates consumption	Higher Na intake associated with increased stroke risk *Vascular death* 500 mg/d increase HR=1.02, CI: 0.95, 1.10 *By quartile* Q1: HR=1.0 Q2: HR=1.39, CI: 0.95, 2.04 Q3: HR=1.37, CI: 0.91, 22.07 Q4: HR=1.49, CI: 0.82, 2.72 *Stroke:* 500 mg/d increase *HR=1.17, CI: 1.07, 1.27 By quartile Q1: HR=1.0 Q2: HR=1.33, CI: 0.81, 2.18 Q3: HR=1.31, CI: 0.78, 2.22 *Q4: HR=2.50, CI: 1.23, 5.07

continued

TABLE F-2 Continued

Citation	Population Studied	Study Design	Sample Size	Sodium Exposure (method and level)
Gardener et al., 2012 continued				
Geleijnse et al., 2007	Randomly selected Dutch Rotterdam Study subjects, >55 y (41% men; mean age=69.2 y)	Population-based prospective cohort	1,448	Overnight urine sample UNa excretion concentration standardized from 24-h values using recorded collection times and urinary volumes Analyses per 1 SD increase in UNa excretion

Follow-up Period	Health Outcome	Confounders Adjusted for	Results
			Stroke, MI, or vascular death: 500 mg/d increase *HR=1.06, CI: 1.00, 1.12
			By quartile Q1: HR=1.0 Q2: HR=1.32, CI: 0.98, 1.78 Q3: HR=1.21, CI: 0.87, 1.67 *Q4: HR=1.70, CI: 1.08, 2.68
			MI: 500 mg/d increase HR=0.95, CI: 0.86, 1.04
			By quartile Q1: HR=1.0 Q2: HR=0.93, CI: 0.58, 1.51 Q3: HR=0.68, CI: 0.40, 1.15 Q4: HR=0.79, CI: 0.37, 1.69
5.5 y (median)	ACM CVD mortality Incident MI Incident stroke	Age, sex, 24-h urinary creatinine excretion, 24-h urinary potassium, BMI, smoking status, DM, use of diuretics, highest completed education, dietary confounders (intake of total energy, alcohol, Ca, sat. fat)	No association between Na intake and mortality UNa excretion not significantly associated with incident MI or stroke *ACM* HR=0.95, CI: 0.81, 1.12 *CVD mortality* HR=0.77, CI: 0.60, 1.01 (borderline significant inverse association was reduced when subjects with a history of CVD/ hypertension were excluded) *Incident MI* HR=1.19, CI: 0.97, 1.46 *Incident stroke* HR=1.08, CI: 0.80, 1.46

continued

TABLE F-2 Continued

Citation	Population Studied	Study Design	Sample Size	Sodium Exposure (method and level)
Heerspink et al., 2012	Subjects from the RENAAL (250 centers in 28 countries in the Americas, Asia, Australia, and Europe) and IDNT trials (210 centers in the Americas, Europe, Israel, and Australia) (intervention was therapy with angiotensin receptor blockers), 30-70 y, with type 2 diabetic nephropathy, proteinuria (>500 mg/d, RENAAL; >900 mg/d, IDNT), and serum creatinine levels 1.3-3.0 mg/dl (RENAAL) or 1.0-30 mg/dl (IDNT) Randomized to ARB vs. non-RAAS therapy	Prospective cohort	1,177 (769 men; 408 women)	Multiple 24-h urine collections Na intake tertiles based on 24-h Na/creatinine ratios: T1: <2,783 mg/d T2: 2,783-3,519 mg/d T3: ≥3,519 mg/d
Jafar, 2006	Pashtun ethnic subgroup in Pakistan, mean age=51.5 y	Prospective cohort	500	FFQ that included a question on use of extra salt on food in addition to salt used in cooking
Kono et al., 2011	Japanese with acute IS who met the criteria for emergent admission to an acute hospital, mean age=63.9±9.1 y	Prospective cohort	102 (78 men; 24 women) Analysis performed on 89 patients	Salt intake measured using a self-monitoring device Urine collected daily for 3 consecutive days for approximately 8 h/d

Follow-up Period	Health Outcome	Confounders Adjusted for	Results
30 mo. 4-wk intervals until 3 mo. Then at 3-mo. intervals	CVD (CVD mortality, MI, stroke, hospitalization for heart failure or revascularization procedure)	None listed	ARBs were significantly more effective at decreasing CVD when Na intake was in the lowest tertile (<2,783 mg/day) T1: HR=0.63 CI: 0.43, 0.92 T2: HR=1.02, CI: 0.73, 1.43 T3: HR=1.25, CI: 0.89, 1.75 *p for interaction=0.021
1 y	Stroke	Not listed	*OR=3.66, CI: 1.06, 12.60
3 y	Recurrence rate for vascular events	Age, medication	Higher salt intake predictive for stroke recurrence *HR=1.98, CI: 1.02, 4.22, p=0.028

continued

TABLE F-2 Continued

Citation	Population Studied	Study Design	Sample Size	Sodium Exposure (method and level)
Larsson et al., 2008	Finnish men in the Alpha-Tocopheral, Beta-Carotene Cancer Prevention (ATBC) Study, 50-69 y who smoked ≤5 cigarettes/d at baseline	Prospective cohort	26,556	Self-administered 276-item FFQ Nutrient intake calculated using food composition database at the National Public Health Institute Na intake quintiles (median) adjusted for energy intake: Q1: 3,909 mg/d Q2: 4,438 mg/d Q3: 4,810 mg/d Q4: 5,212 mg/d Q5: 5,848 mg/d
Nagata et al., 2004	Nonhospitalized Japanese in Takayama City, Gifu, ≥35 y Exclude people who reported having stroke, IHD, or cancer	Population-based prospective cohort	29,079 (13,355 men; 15,724 women)	Semi-quantitative 169-item FFQ Na intake estimated from *Standard Tables of Food Composition in Japan* Na intake tertiles: Men: T1: 4,070 mg/d T2: 5,209 mg/d T3: 6,613 mg/d Women: T1: 3,799 mg/d T2: 4,801 mg/d T3: 5,930 mg/d

Follow-up Period	Health Outcome	Confounders Adjusted for	Results
13.6 y (mean)	Cerebral infarction Intracerebral hemorrhage Subarachnoid hemorrhage	Age, supplementation group, number of cigarettes/d, BMI, systolic and diastolic BP, serum total cholesterol, serum HDL, history of diabetes, history of CHD, leisure time physical activity, alcohol intake, total energy intake	Na intake not significantly associated with stroke or subarachnoid hemorrhage *Cerebral infarction* Q1: RR=1.00 Q2: RR=1.08, CI: 0.96, 1.22 Q3: RR=1.05, CI: 0.93, 1.18 Q4: RR=0.99, CI: 0.87, 1.13 Q5: RR=1.04, CI: 0.92, 1.18 p for trend=0.99 *Intracerebral hemorrhage* Q1: RR=1.00 Q2: RR=0.81, CI: 0.58, 1.13 Q3: RR=0.99, CI: 0.71, 1.37 Q4: RR=1.04, CI: 0.75, 1.44 Q5: RR=1.28, CI: 0.93, 1.75 p for trend=0.06 *Subarachnoid hemorrhage* Q1: RR=1.00 Q2: RR=0.69, CI: 0.44, 1.08 Q3: RR=0.81, CI: 0.52, 1.24 Q4: RR=0.74, CI: 0.48, 1.16 Q5: RR=0.84, CI: 0.54, 1.30 p for trend=0.55
7 y	Stroke mortality	Age, level of education, marital status, BMI, smoking status, alcohol consumption, histories of diabetes and hypertension, energy	Increased Na intake associated with increased risk of stroke mortality *Men* T2: HR=1.60, CI: 0.92, 2.80 T3: HR=2.33, CI: 1.23, 4.45 *p for trend: 0.009 *Women* T2: HR=1.33, CI: 0.80, 2.21 T3: HR=1.70, CI: 0.96, 3.02 p for trend: 0.07

continued

TABLE F-2 Continued

Citation	Population Studied	Study Design	Sample Size	Sodium Exposure (method and level)
O'Donnell et al., 2011	Participants in the ONTARGET and TRANSCEND trials, at high risk of CVD (with CVD or DM), ≥55 y (mean age=66.52 y) Recruited from 733 centers in 40 countries Ineligible if they had CHF, low ejection fraction, significant valvular disease, serum creatinine >3.0 mg/L, renal artery stenosis, nephritic range proteinuria, BP >160/100 mmHG	Follow-up of two RCT cohorts treated with ACE inhibitor or AGII antagonist	28,880 (20,376 men; 8,504 women)	Morning fasting urine sample used to estimate 24-h Na excretion using the Kawasaki formula 7 Na excretion levels: 1: <2,000 mg/d 2: 2,000-2,999 mg/d 3: 3,000-3,999 mg/d 4: 4,000-5,999 mg/d 5: 6,000-6,999 mg/d 6: 7,000-8,000 mg/d 7: >8,000 mg/d

Follow-up Period	Health Outcome	Confounders Adjusted for	Results
56 mo. (median)	ACM CVD mortality Non-CVD mortality Stroke MI CHF hospitalization	Age, sex, race/ethnicity, prior stroke or MI, creatinine, BMI, hypertension, DM, atrial fibrillation, smoking, LDL, HDL, treatment allocation (with ramipril, telmitarsan or both, statins, β-blockers, diuretics, Ca antagonist, antithrombotic therapy), fruit and vegetable consumption, level of exercise, UNa and UK excretion, baseline BP, changes in systolic BP from baseline to last follow-up	J-shaped association between Na excretion and CVD mortality Higher Na intake associated with increased risk of stroke, MI, and CHF hospitalization. Lower Na intake associated with increased risk of CHF hospitalization. *ACM* 1: HR=1.19, CI: 0.99, 1.45 2: HR=1.11, CI: 0.99, 1.26 3: HR=1.06, CI: 0.96, 1.16 4: HR=1.00 *5: HR=1.14, CI: 1.02, 1.28 *6: HR=1.29, CI: 1.10, 1.52 *7: HR=1.56, CI: 1.30, 1.89 *CVD mortality* *1: HR=1.37, CI: 1.09, 1.73 *2: HR=1.19, CI: 1.02, 1.39 3: HR=1.09, CI: 0.96, 1.23 4: HR=1.00 5: HR=1.11, CI: 0.96, 1.29 *6: HR=1.53, CI: 1.26, 1.86 *7: HR=1.66, CI: 1.31, 2.10 *Non-CVD mortality* 1: HR=0.92, CI: 0.65, 1.29 2: HR=1.00, CI: 0.83, 1.21 3: HR=1.02, CI: 0.88, 1.18 4: HR=1.00 5: HR=1.18, CI: 0.99, 1.40 6: HR=0.95, CI: 0.71, 1.27 *7: HR=1.42, CI: 1.04, 1.94 *Stroke* 1: HR=1.06, CI: 0.76, 1.46 2: HR=1.05, CI: 0.86, 1.28 3: HR=0.97, CI: 0.83, 1.13 4: HR=1.00 5: HR=0.95, CI: 0.79, 1.15 6: HR=1.06, CI: 0.81, 1.40 *7: HR=1.48, CI: 1.09, 2.01 *MI* 1: HR=1.10, CI: 0.80, −1.53 2: HR=1.04, CI: 0.85, 1.27 3: HR=1.11, CI: 0.96, 1.28 4: HR=1.00 *5: HR=1.21, CI: 1.03, 1.43 6: R=1.11, CI: 0.85, 1.44 *7: HR=1.48, CI: 1.11, 1.98

continued

TABLE F-2 Continued

Citation	Population Studied	Study Design	Sample Size	Sodium Exposure (method and level)
O'Donnell et al., 2011 continued				
Stolarz-Skrzypek et al., 2011	Individuals >20 y invited to participate in 2 studies: (1) Flemish Study on Environment, Genes, and Health Outcome; (2) European Project on Genes in Hypertension	Population-based prospective cohort	3,681	24-h urine collection Na excretion tertiles: T1: 2,461 mg/d T2: 3,864 mg/d T3: 5,980 mg/d

Follow-up Period	Health Outcome	Confounders Adjusted for	Results
			CHF hospitalization 1: HR=1.29, CI:0.95, 1.74 *2: HR=1.23, CI: 1.01, 1.49 3: HR=1.07, CI: 0.91, 1.25 4: HR=1.00 5: HR=1.04, CI: 0.79, 1.42 6: HR=1.06, CI: 0.79, 1.42 *7: HR=1.51, CI: 1.12, 2.05
7.9 y (median)	ACM CVD mortality Noncardiovascular mortality Fatal and nonfatal CVD Fatal and nonfatal coronary Fatal and nonfatal stroke	Study population, sex, age, BMI, systolic BP, 24-h UK excretion, antihypertensive drug treatment, smoking, alcohol, diabetes, total cholesterol, educational attainment	Lower Na intake associated with higher CVD mortality Na intake not significantly associated with CVD events *ACM* T1: HR=1.14, CI: 0.87, 1.50 T2: HR=0.94, CI: 0.75, 1.18 T3: HR=1.06, CI: 0.84, 1.33 p for trend=0.10 *CVD mortality* T1: HR=1.56, CI: 1.02, 2.36 T2: HR=1.05, CI: 0.72, 1.53 T3: HR=0.95, CI: 0.66, 1.38 *p for trend=0.02 *Noncardiovascular mortality* T1: HR=0.98, CI: 0.71, 1.36 T2: HR=0.90, CI: 0.68, 1.20 T3: HR=1.11, CI: 0.83, 1.47 p for trend=0.64 *Fatal and nonfatal CVD* T1: HR=1.13, CI: 0.90, 1.42 T2: HR=1.11, CI: 0.90, 1.36 T3: HR=0.90, CI: 0.73, 1.11 p for trend=0.55 *Fatal and nonfatal coronary* T1: HR=1.42, CI: 0.99, 2.04 T2: HR=1.17, CI: 0.89, 1.54 T3: HR=0.86, CI: 0.65, 1.13 p for trend=0.10 *Fatal and nonfatal stroke* T1: HR=1.07, CI: 0.57, 2.00 T2: HR=1.29, CI: 0.75, 2.20 T3: HR=0.78, CI: 0.45, 1.33 p for trend=0.64

continued

TABLE F-2 Continued

Citation	Population Studied	Study Design	Sample Size	Sodium Exposure (method and level)
Takachi et al., 2010	Japanese subjects in the two cohorts of the Japan Public Health Center-based Prospective Study 40-59 y (cohort I) and 40-69 y (cohort II)	Prospective cohort	77,500 (35,730 men; 41,770 women)	138-item FFQ Na intake calculated using the *Standardized Tables of Food Composition, 5th edition revised* Validated with 24-h UNa excretion in subsamples Na intake quintiles based on median intake Q1: 3,084 mg/d Q2: 4,005 mg/d Q3: 4,709 mg/d Q4: 5,503 mg/d Q5: 6,844 mg/d
Thomas et al., 2011	Finnish, diagnosed with type 1 diabetes diagnosed before 35 y, without ESRD at baseline Mean age=39 y; median duration of diabetes= 20 y	Prospective cohort	2,807	Single 24-h urine collection Na excretion tertiles T1: <2,346 mg/d T2: 2,346-4,301 mg/d T3: >4,301 mg/d

Follow-up Period	Health Outcome	Confounders Adjusted for	Results
10 y (cohort I)	CVD	Sex, age, BMI, smoking status, alcohol consumption, physical activity, quintiles of energy, K, and Ca	Higher Na intake associated with increased risk of CVD, significant increase in stroke, but not MI
	Stroke		
7 y (cohort II)	MI		*CVD*
			Q1: HR=1.00
			Q2: HR=1.11, CI: 0.96, 1.29
			Q3: HR=1.02, CI: 0.87, 1.19
			Q4: HR=1.10, CI: 0.94, 1.29
			Q5: HR=1.19, CI: 1.01, 1.40
			p for trend=0.06
			Stroke
			Q1: HR=1.00
			Q2: HR=1.05, CI: 0.90, 1.24
			Q3: HR=0.97, CI: 0.82, 1.14
			Q4: HR=1.08, CI: 0.92, 1.28
			Q5: HR=1.21, CI: 1.01, 1.43
			*p for trend=0.03
			MI
			Q1: HR=1.00
			Q2: HR=1.50, CI: 1.05, 2.14
			Q3: HR=1.34, CI: 0.92, 1.96
			Q4: HR=1.26, CI: 0.85, 1.88
			Q5: HR=1.09, CI: 0.71, 1.68
			p for trend=0.91
10 y (median)	ACM	Age, sex, duration of diabetes, presence/ severity of CKD, presence of established CVD, systolic BP	UNa excretion significantly associated with ACM (*p<0.001).
			T1 and T3 had reduced cumulative survival

continued

TABLE F-2 Continued

Citation	Population Studied	Study Design	Sample Size	Sodium Exposure (method and level)
Tikellis et al., 2013	Finnish Diabetic Nephropathy Study subjects with type 1 diabetes (onset before 40 y) without prior CVD or ESRD Subpopulation of Thomas et al., 2011 (above) excluding those with prior CVD	Prospective cohort	2,648 Subset of 2,807 included in Thomas et al., 2011 (above)	24-h urine collection at baseline Na excretion quartiles T1: <2,346 mg/d T2: 2,346-4,301 mg/d T3: >4,301 mg/d
Umesawa et al., 2008	Japanese subjects in the Japan Collaborative Study for Evaluation of Cancer Risk, 40-79 y with no history of stroke, CHD, or cancer	Prospective cohort; mortality follow-up on population study	58,730 (23,119 men; 35,611 women)	35-item FFQ Responses were *Rarely *1-2 d/mo *1-2 d/wk *3-4 d/wk *Almost every day Based on results of validation study comparing FFQ to four 3-d dietary records for 85 people, intake values calibrated by multiplying by 2 Calibrated Na intake quintiles Q1: 2,323 mg/d Q2: 3,358 mg/d Q3: 4,186 mg/d Q4: 5,060 mg/d Q5: 6,256 mg/d

Follow-up Period	Health Outcome	Confounders Adjusted for	Results
10 y (median)	ACM CVD	Parameters associated with daily UNa excretion, age, sex, glycemic control, presence/severity of CKD, lipid levels	After adjustment, UNa excretion significantly associated with increased risk of ACM and incidence CVD event Nonlinear association, individuals with highest and lowest Na excretion had reduced cumulative survival Na excretion not significantly associated with stroke or new coronary event (Rates and p values not provided)
12.7 y (mean)	Total stroke mortality CVD mortality CHD mortality	Age, sex, BMI, smoking status, alcohol intake, history of hypertension, history of diabetes, menopause, hormone replacement therapy, time spent on sports activities, walking time, educational status, perceived mental stress, Ca intake, K intake (quintiles)	Higher Na intake associated with increased risk of stroke and CVD mortality *Stroke mortality* Q1: HR=1.00 Q2: HR=0.96, CI: 0.76, 1.22 Q3: HR=1.26, CI: 1.00, 1.59 Q4: HR=1.42, CI: 1.12, 1.80 Q5: HR=1.55, CI: 1.21, 2.00 *p for trend<0.001 *CVD mortality* Q1: HR=1.00) Q2: HR=1.04, CI: 0.89, 1.22 Q3: HR=1.19, CI: 1.01, 1.39 Q4: HR=1.29, CI: 1.10, 1.52 Q5: HR=1.42, CI: 1.20, 1.69 *p for trend<0.001 *CHD mortality* Q1: HR=1.00 Q2: HR=0.92, CI: 0.66, 1.28 Q3: HR=1.05, CI: 0.75, 1.46 Q4: HR=1.09, CI: 0.77, 1.54 Q5: HR=1.19, CI: 0.82, 1.73 p for trend=0.230

continued

TABLE F-2 Continued

Citation	Population Studied	Study Design	Sample Size	Sodium Exposure (method and level)
Yang et al., 2011	NHANES III, 76% <60 y	Prospective cohort	12,267 (5,899 men; 6,368 women)	24-h dietary recall of Na intake Na intake quartiles by midvalue of quartile of estimated usual intake in population: Q1: 2,176 mg/d Q2: 3,040 mg/d Q3: 3,864 mg/d Q4: 5,135 mg/d

Follow-up Period	Health Outcome	Confounders Adjusted for	Results
14.8 y (mean)	ACM CVD mortality IHD mortality	Sex, race/ethnicity, educational attainment, BMI, smoking status, alcohol intake, total cholesterol, HDL cholesterol, physical activity, family history of CVD, total calorie intake	Higher Na intake associated with increased ACM *ACM* Q1: HR=1.00 Q2: HR=1.17, CI: 1.13, 1.33 Q3: HR=1.37, CI: 1.28, 1.74 Q4: HR=1.73, CI: 1.54, 2.63 *p for trend=0.02 *CVD mortality* Q1: HR=1.00 Q2: HR=0.95, CI: 0.71, 1.27 Q3: HR=0.90, CI: 0.51, 1.60 Q4: HR=0.83, CI: 0.31, 2.28 p for trend=0.72 *IHD mortality* Q1: HR=1.00 Q2: HR=1.17, CI: 0.84-1.62 Q3: HR=1.36, CI: 0.71, 2.58 Q4: HR=1.70, CI: 0.55, 5.27 p for trend=0.36

TABLE F-3 Evidence Tables: CVD/Stroke/Mortality Case-Control Studies

Citation	Population Studied	Study Design	Sample Size (case/ control)
Baune et al., 2005	Cases: Patients in the Gaza Strip who had been hospitalized for acute stroke and history of hypertension, 40-69 y, 52% men Controls: Patients in the Gaza Strip with hypertension and no history of stroke, 40-69 y, 52% men	Hospital-based case-control	112 cases 224 controls

Sodium Exposure (method and level)	Health Outcome	Confounders Adjusted for	Results
Questionnaire including 1 question on "excessive use of salt" (yes/no)	Stroke	Age, sex Significant differences in education not controlled for	Significant association between stroke and excessive use of salt at meals *OR=4.51, CI: 2.05-9.90, p=0.0002

TABLE F-4 Evidence Tables: Congestive Heart Failure Randomized Controlled Trials

Citation	Population Studied	Intervention/ Control	Sample Size	Sodium Exposure (method and level)
Parrinello et al., 2009	Italian decompensated CHF patients (NYHA class II), 53-86 y LVEF <35%; Serum creatinine <2 mg/dl; blood urea nitrogen ≤60 mg/dL, urinary volume <500 mL/24 h; low natriuresis (<60 mEq/24 h)	*Intervention* 1,840 mg/d Na *Control* 2,760 mg/d Na	*Intervention* 87 (53 men; 34 women) *Control* 86 (56 men; 30 women)	*Intervention* Diets containing 1,840 mg Na *Control* Intervention group diet + 920 mg/d Na (same amt of sat fat, fruit, etc.) Diaries to record fluid intake and diet variations
Paterna et al., 2008	Italian compensated CHF patients (NYHA class II to IV), 53-86 y Ejection fraction <35% Serum creatinine <2 mg/dl	*Intervention* 1,840 mg/d Na *Control* 2,760 mg/d Na	*Intervention* 114 (71 men; 43 women) *Control* 118 (73 men; 45 women)	*Intervention* Diets containing 1,840 mg Na *Control* Intervention group diet + 920 mg/d Na (same amt of sat fat, fruit, etc.) Diaries to record fluid intake and diet variations

Co-intervention	Blinding	Follow-up Period	Health Outcome	Results
1,000 ml/d fluid + 125 or 250 mg furosemide twice a day	Double blind	12 mo Weekly (after 30 d post-discharge) for the first mo, every 2 wks for the next 2 mo, then every mo for the remainder of the study period Length of intervention= 180 d	Mortality Readmissions for worsening CHF	Significantly fewer deaths and readmissions in control group *Mortality* *ARR=14.2, CI: 5.65, 22.7, p<0.005 *Readmissions* *ARR=21.2, CI: 10.8, 31.6, p<0.001
1,000 or 2,000 ml/d fluid + 125 or 250 mg furosemide twice a day	Evaluations by two physicians blinded to the study	Weekly (after 30 d postdischarge) for the first mo, every 2 wks for the next 2 mo, then every mo for the remainder of the study period Length of intervention and follow-up period=180 d	Mortality Readmission for worsening CHF	Fewer deaths and readmissions in control *Mortality* ARR=8.07%, CI: 0.71, 15.43%, p=NS *Readmissions* *ARR=18.69%, CI: 9.29, 28.08%, p<0.05

continued

TABLE F-4 Continued

Citation	Population Studied	Intervention/ Control	Sample Size	Sodium Exposure (method and level)
Paterna et al., 2009	Italian compensated CHF patients (NYHA class II to IV), 55-83 y Ejection fraction <35% Serum creatinine <2 mg/dl	*Intervention* 1,840 mg/d Na (low) *Control* 2,760 mg/d Na (normal)	*Intervention* 205 (77 men; 128 women) *Control:* 205 (75 men; 130 women)	*Intervention* Diets containing 1,840 mg Na *Control* Intervention group diet + 920 mg/d Na (same amt of sat fat, fruit, etc.)
Paterna et al., 2011	Italian compensated CHF patients (NYHA class III to IV), 53-86 y Ejection fraction <40% Serum creatinine <2.5 mg/dl	*Intervention* 1,840 mg/d Na *Control* 2,760 mg/d Na	1,771 patients, 881 (intervention) and 890 (control)	*Intervention* Diets containing 1,840 mg Na *Control* Intervention group diet + 920 mg/d Na (same amt of sat fat, fruit, etc.) Diaries to record fluid intake and diet variations

Co-intervention	Blinding	Follow-up Period	Health Outcome	Results
1,000 ml/d fluid + 250-500 mg/d furosemide twice a day	Evaluations by two physicians blinded to the study	Weekly (after 30 d post-discharge) for the first mo, every 2 wks for the next 2 mo, then every mo for the remainder of the study period Length of intervention and follow-up period=180 d	Readmission for worsening CHF	Normal Na diet associated with significantly reduced readmissions *OR=2.46, CI: 1.84, 3.29, p<0.0001
Intervention Furosemide (250 mg) plus HSS (150 ml) twice daily and fluid intake of 1,000 ml/d *Control* Furosemide (250 mg) twice a day, without HSS and fluid intake of 1,000 ml/d	Evaluations by two physicians blinded to the study	57 mo (mean) Treatment in both groups continued during follow-up	Mortality Hospitalization time Readmission for worsening CHF	*Intervention vs. control* Significant reduction in hospitalization time, readmission rates, and mortality *Mortality* *12.9 vs. 23.8 percent, p<0.0001 *Hospitalization time* *3.5 vs. 5.5 days, p≤0.0001 *Readmissions* *18.5 vs. 34.2 percent, p<0.0001

TABLE F-5 Evidence Tables: Congestive Heart Failure Cohort Studies

Citation	Population Studied	Study Design	Sample Size	Sodium Exposure (method and level)
Arcand et al., 2011	Canadian medically stable, ambulatory CHF patients from two outpatient clinics, mean age 60±13 y	Prospective cohort	123	Two 3-d food records (1st at study entry, 2nd 6-12 wks later) Records analyzed using ESHA Food Processor SQL vs. 10.1 Validated with 2 urine collections in subgroup Na intake tertiles: T1: ≤1,900 mg/d T2: 2,000-2,700 mg/d T3: ≥2,800 mg/d
Lennie et al., 2011	Chronic CHF patients from outpatient clinics in Kentucky, Georgia, Indiana, and Ohio with LVEF <40% or preserved LVEF ≥40%, on stable doses of medication for 3 mo, average age=62±12 y	Prospective cohort	302 (203 men; 99 women)	24-h urine collection UNa excretion levels ≥3,000 mg/d <3,000 mg/d

Follow-up Period	Health Outcome	Confounders Adjusted for	Results
3 y (median)	Mortality or transplantation ADHF events All-cause hospitalization	Age, sex, caloric intake, LVEF, BMI, furosemide use, β-blockers use	Higher Na intake associated with mortality, ADHF, and all-cause hospitalization *Mortality* *T3 vs. T1* *HR=3.54, CI: 1.46, 8.62, p=0.005 *ADHF* *T3 vs. T1* *HR=2.55, CI: 1.61, 4.04, p<0.001 *All-cause hospitalization* *T3 vs. T1* *HR=1.39, CI: 1.06, 1.83, p=0.018
12 mo	Event-free survival	Age, sex, CHF etiology, BMI, ejection fraction, total comorbidity score	Higher UNa excretion levels (≥3,000 mg/d) significantly associated with longer event-free survival *NYHA class I/II* *HR=0.44, CI: 0.20, 0.97, p=0.040 *NYHA class III/IV* *HR=2.54, CI: 1.10, 5.83, p=0.028

TABLE F-6 Evidence Tables: Kidney Disease Cohort Studies

Citation	Population Studied	Study Design	Sample Size	Sodium Exposure
Heerspink et al., 2012	Subjects from the RENAAL (250 centers in 28 countries in the Americas, Asia, and Europe) and IDNT trials 209 centers in the Americas, Australia, Europe, and Israel) (intervention was therapy with angiotensin receptor blockers), 30-70 y, with type 2 diabetic nephropathy, proteinuria (>500 mg/d, RENAAL; >900 mg/d, IDNT), and serum creatinine levels 1.3-3.0 mg/dl (RENAAL) or 1.0-30 mg/dl (IDNT) Randomized to ARB vs. non-RAAS therapy	Prospective cohort	1,177 (769 men; 408 women)	Multiple 24-h urine collections Na intake tertiles based on 24-h Na/creatinine ratios: T1: <2,783 mg/d T2: 2,783-3,519 mg/d T3: ≥3,519 mg/d
McCausland et al., 2012	Hemodialysis Study subjects (United States), mean age=58±14 y, 44% men, 63% African American, 44% diabetic	Post-hoc analysis of a prospective cohort	1,770	2-d diet diary assisted calls Restricted cubic spline, knots at 1,500, 2,000, and 2,500 mg/d Na intake quartiles

Follow-up Period	Health Outcome	Confounders Adjusted for	Results
30 mo 4-wk intervals until 3 mo, then at 3-mo intervals	CKD progression (doubling of serum creatinine or incident ESRD) ESRD		ARBs were significantly more effective at decreasing CKD progression when Na was in the lowest tertile (<2,783 mg/d) *CKD progression* T1: HR=0.57 CI: 0.39, 0.84 T2: HR=1.00, CI: 0.70, 1.42 T3: HR=1.37, CI: 0.96, 1.96 *p for interaction<0.001 *ESRD* T1: HR=0.54 CI: 0.34, 0.86 T2: HR=0.82, CI: 0.54, 1.26 T3: HR=1.35, CI: 0.88, 2.07 *p for interaction=0.005
2.1 y (median)	ACM	Age, sex, race (African American vs. non-African American), Hemodialysis Study Kt/V and flux group assignments, post-dialysis weight, sex-by-weight cross-product term access, CHF status, presence absence of diabetes and IHD	Significant association between higher Na intake and increased risk of death (rates not provided)

continued

TABLE F-6 Continued

Citation	Population Studied	Study Design	Sample Size	Sodium Exposure
Thomas et al., 2011	Finnish, diagnosed with type 1 diabetes diagnosed before 35 y, without ESRD at baseline Mean age=39 y; median duration of diabetes= 20 y	Prospective cohort	2,807	Single 24-h urine collection Na excretion tertiles T1: <2,346 mg/d T2: 2,346-4,301 mg/d T3: >4,301 mg/d

Follow-up Period	Health Outcome	Confounders Adjusted for	Results
10 y (median)	ESRD	Age, sex, duration of diabetes, presence/severity of CKD, presence of established CVD, systolic BP	UNa excretion significantly associated with ESRD (*p<0.001) T1 had the highest cumulative incidence of ESRD

TABLE F-7 Evidence Tables: Diabetes Cohort Studies

Citation	Population Studied	Study Design	Sample Size	Sodium Exposure (method and level)
Hu et al., 2005	Finnish, 35-64 y	Prospective cohort	1,935 (932 men; 1,003 women)	Self-administered questionnaire that included questions on *Type of food usually consumed *Amount of food consumed *Frequency of consumption of vegetables, fruit, and sausages 24-h urine collection 24-h UNa excretion quartile cutpoints: *Men (in 1982 sample)* 3,795 mg/d 4,876 mg/d 6,210 mg/d *Men (in 1987 sample)* 3,496 mg/d 4,646 mg/d 5,819 mg/d *Women (in 1982 sample)* 2,806 mg/d 3,657 mg/d 4,600 mg/d *Women (in 1987 sample)* 2,691 mg/d 3,450 mg/d 4,347 mg/d

Follow-up Period	Health Outcome	Confounders Adjusted for	Results
18.1 y (mean)	Type 2 diabetes incidence	Age, sex, study year, BMI, physical activity, systolic BP, antihypertensive drug treatment, education, smoking, coffee, alcohol, fruit, vegetable, sausage, bread, and sat. fat consumption	Higher Na intake associated with increased risk of type 2 diabetes Q4 vs. Q1 Q1-Q3: HR=1.00 *Q4: HR=2.05, CI: 1.43, 2.96

continued

TABLE F-7 Continued

Citation	Population Studied	Study Design	Sample Size	Sodium Exposure (method and level)
Roy and Janal, 2010	African Americans in New Jersey with type 1 diabetes Mean ages: men, 26.7 y; women, 27.8 y	Prospective cohort	469	Reduced 60-item FFQ (BRIEF87) of the Health Habits and History Questionnaire developed by NCI; administered by a research assistant Recorded average frequency of consumption and serving sizes; nutrient intakes calculated using DietSys version 3.0 and NCI nutrient databases

Follow-up Period	Health Outcome	Confounders Adjusted for	Results
6 y	Macular edema (ME)	Baseline age, sex, glycated hemoglobin level, hypertension, proteinuria, blood cholesterol level, socioeconomic status, physical exercise, calories	Baseline Na intake significantly, positively associated with incidence of ME *ME* *OR=1.43, CI: 1.10, 1.86, p=0.08

TABLE F-8 Evidence Tables: Metabolic Syndrome and Diabetes
Cross-Sectional Studies

Citation	Population Studied	Study Design	Sample Size
Daimon et al., 2008	Japanese subjects in the Takahata study, >35 y	Population-based cross-sectional	2,956
Rodrigues et al., 2009	Patients, 25-64 y, who went to the University Hospital in Brazil to undergo clinical and laboratory exams	Population-based cross-sectional	1,662

Sodium Exposure (method and level)	Health Outcome	Confounders Adjusted for	Results
Brief diet history recall questionnaire to record dietary habits over a 1-month period and measure salt consumption	Role of genetic polymorphism linked to salt intake in diabetes risk	Age, sex, BMI, serum remnant-like particle cholesterol	Significant association between genetic polymorphism with diabetes in subjects with salt intake <12,440 mg/d (*p=0.032)
Salt (sodium) intake levels ≥12,440 (5,376) mg/d <12,440 (5,376) mg/d			
12-h nocturnal urine collection Daily Na intake estimated based on 45% of total daily Na excreted at night	Metabolic syndrome components (waist circumference, triglycerides level, HDL cholesterol level, glucose level)	Weight (waist circumference)	No significant association between UNa excretion and metabolic syndrome components when normotensive individuals were stratified by sex and number of metabolic syndrome components (p=0.49 for men, p=0.63 for women)

continued

TABLE F-8 Continued

Citation	Population Studied	Study Design	Sample Size
Teramoto et al., 2011	Participants in the Olmesartan Mega Study to Determine the Relationship between Cardiovascular Endpoints and Blood Pressure Goal Achievement (OMEGA); olmesartan-naïve Japanese adults, 50-79 y diagnosed with hypertension receiving treatment at outpatient clinics	Prospective cohort	9,585 8,576 at follow-up

Sodium Exposure (method and level)	Health Outcome	Confounders Adjusted for	Results
FFQ, including questions on consumption of high-salt foods Measured as *no intake *1-2/wk *3-5/wk *intake every day Estimated Na intake calculated using formula from Arakawa et al. (2009) Divided into Na intake quartiles (not given) and then 2 score groups *<20 (greater than 75th percentile of intake) *≥20	Metabolic syndrome	Age	The highest quartile of Na intake was associated with higher prevalence of metabolic syndrome in men but not in women *p=0.0026

TABLE F-9 Evidence Tables: Gastrointestinal Cancer Cohort Studies

Citation	Population Studied	Study Design	Sample Size	Sodium Exposure (method and level)
Murata et al., 2010	Japanese, 40-79 y, without a cancer diagnosis	Population cohort	6,830 (3,074 men; 3,756 women)	Self-administered dietary questionnaire to assess usual intake of salted foods (e.g., 1/day, 2-4/week) Classified as "high intake" or "low intake"

Follow-up Period	Health Outcome	Confounders Adjusted for	Results
13.9 y	Stomach cancer mortality	Age, BMI, physical activity, smoking, alcohol, history of diabetes, intake of vegetables, fruit, tea, red meat, processed meat	Higher Na intake associated with increased risk of stomach and rectal cancer in men, but not women
	Rectal cancer mortality		
	Esophageal cancer mortality		*High vs. Low intake* *Stomach cancer* *Men: OR=2.05, CI: 1.25, 3.38, p<0.05 Women: OR=1.93, CI: 0.87, 4.88, p=NS
	Colon cancer mortality		
			Rectal cancer *Men: HR=3.58, CI: 1.08, 11.9, p<0.05 Women: HR=0.40, CI: 0.05, 3.47, p=NS
			Colon cancer Men: HR=1.43, CI: 0.66, 3.67, P=NS Women: HR=2.21, CI: 0.63, 7.78, p=NS
			Esophageal cancer Men: HR=1.55, CI: 0.18, 6.39, p=NS Women: HR=1.22, CI: 0.35, 5.43, p=NS

continued

TABLE F-9 Continued

Citation	Population Studied	Study Design	Sample Size	Sodium Exposure (method and level)
Shikata et al., 2006	Japanese, ≥40 y, with no history of gastrectomy or gastric cancer Mean ages: 57.3 y (men), 58.7 y (women)	Prospective cohort	2, 467 (1,023 men; 1,444 women)	70-item self-administered FFQ over the last year Nutritional intake was calculated using the 4th revision of the *Standard Tables of Food Composition in Japan* Adjusted for energy intake with Willet and Stamper method Salt (sodium) intake quartiles: Q1: <10,000 (4,000) mg/d Q2: 10,000 (4,000)-12,900 (5,160) mg/d Q3: 13,000 (5,200)-15,900 (6,360) mg/d Q4: ≥16,000 (6,400) mg/d
Sjödahl et al., 2008	Norwegian (mean age at baseline: 49 y)	Population-based, prospective cohort	73,133 (35,955 men; 37,178 women)	FFQ: average frequency of dietary intake of salted foods (never or <1/mo, 1-2/mo, up to 1/wk, up to 2/wk, more than 2/wk) with no list of foods to select Assessed frequency of intake of salted foods and sprinkling extra salt on food, and then calculated a summary score of salt intake

Follow-up Period	Health Outcome	Confounders Adjusted for	Results
14 y	Gastric cancer	Age, sex, *H. pylori* infection, atrophic gastritis, medical history of peptic ulcer, family history of cancer, BMI, diabetes, cholesterol, physical activity, alcohol, smoking, intake of total energy, protein, carbohydrate, dietary fiber, and vitamins B_1, B_2, C	Positive association between dietary salt and gastric cancer compared to Q1 *Q2: HR=2.12, CI: 1.08, 4.17, p<0.05 Q3: HR=1.88, CI: 0.91, 3.89, NS *Q4: HR=2.67, CI: 1.36, 5.24, p<0.01 Significant association between gastric cancer and atrophic gastritis + *H. pylori* infection *HR=2.87, CI: 1.14, 7.24, p<0.05
15.4 y	Gastric adenocarcin-oma	Age, gender, smoking status, alcohol use, physical activity, occupation level	No statistically significant association between levels of intake of salted foods and risk of gastric adenocarcinoma *Intake of salted foods* p for trend=0.39 *Sprinkling extra salt on food* p for trend=0.56 *Summary score of salt intake* p for trend=0.87

continued

TABLE F-9 Continued

Citation	Population Studied	Study Design	Sample Size	Sodium Exposure (method and level)
Takachi et al., 2010	Japanese subjects in the 2 cohorts of the Japan Public Health Center-based Prospective Study 40-59 y (cohort I) and 40-69 y (cohort II)	Prospective cohort	77,500 (35,730 men; 41,770 women)	138-item FFQ Na intake calculated using the *Standardized Tables of Food Composition, 5th edition revised* Validated with 24-h UNa excretion in subsamples Na intake quintiles (median): Q1: 3,084 mg/d Q2: 4,005 mg/d Q3: 4,709 mg/d Q4: 5,503 mg/d Q5: 6,844 mg/d

Follow-up Period	Health Outcome	Confounders Adjusted for	Results
10 y (cohort I)	Total cancer	Sex, age, BMI, smoking status, alcohol consumption, physical activity, quintiles of energy, K, and Ca	Higher Na consumption not associated with increased risk of cancer
7 y (cohort II)	Gastric cancer		*Total cancer* Q1: HR=1.00 Q2: HR=1.02, CI: 0.93, 1.13 Q3: HR=1.07, CI: 0.96, 1.18 Q4: HR=1.01, CI: 0.91, 1.12 Q5: HR=1.04, CI: 0.93, 1.16 p for trend=0.61
	Colorectal cancer		*Gastric cancer* Q1: HR=1.00 Q2: HR=1.05, CI: 0.84, 1.31 Q3: HR=1.06, CI: 0.84, 1.34 Q4: HR=1.05, CI: 0.83, 1.34 Q5: HR=1.07, CI: 0.83, 1.38 p for trend=0.64
			Colorectal cancer Q1: HR=1.00 Q2: HR=1.05, CI: 0.84, 1.33 Q3: HR=1.08, CI: 0.85, 1.37 Q4: HR=1.08, CI: 0.84, 1.37 Q5: HR=1.10, CI: 0.85, 1.42 p for trend=0.51

continued

TABLE F-9 Continued

Citation	Population Studied	Study Design	Sample Size	Sodium Exposure (method and level)
Tsugane et al., 2004	Japanese, 40-59 y, without a self-reported serious illness (cancer, cerebrovascular disease, MI, chronic liver disease) Individuals were from 14 administrative districts supervised by 4 regional public health centers	Population-based prospective cohort	39,065 (18,684 men; 20,381 women)	Self-administered 27-item FFQ assessing weekly intake Na intake calculated using the *Standardized Tables of Food Composition* (Science and Technology Agency, 1982) Individuals with extreme energy intakes were excluded (upper and lower 2.5%) Validated with 28-day dietary record Na intake quintiles by median: Q1: 2,900 mg/day Q2: 4,800 mg/day Q3: 6,100 mg/day Q4: 7,500 mg/day Q5: 9,900 mg/day

Follow-up Period	Health Outcome	Confounders Adjusted for	Results
12 y	Gastric cancer	Age, smoking, fruit and non-green-yellow vegetable intake	High salted foods were strongly associated with gastric cancer in men *Men* Q1: RR=1.00 Q2: RR=1.74, CI: 1.14, 2.66 Q3: RR=1.96, CI: 1.30, 2.97 Q4: RR=2.30, CI: 1.53, 3.46 Q5: RR=2.23, CI: 1.48, 3.35 *p for trend<0.001 *Women* Q1: RR=1.00 Q2: RR=0.86, CI: 0.47, 1.56 Q3: RR=0.96, CI: 0.54, 1.72 Q4: RR=0.58, CI: 0.30, 1.12 Q5: RR=1.32, CI: 0.76, 2.28 p for trend=0.48 Further stratification by study location diminished the association

continued

TABLE F-9 Continued

Citation	Population Studied	Study Design	Sample Size	Sodium Exposure (method and level)
van den Brandt et al., 2003	Dutch, 55-69 y, excluding those with stomach cancer at baseline	Prospective cohort	120,852 (58,279 men; 62,573 women)	Semiquantitative 150-item FFQ (dietary salt intake, salty food intake, added salt) Dietary Na calculated using computerized Dutch food composition table and validated against 9 dietary records Na intake adjusted for energy intake Na intake quintiles by median: Q1: 1,640 mg/d Q2: 2,040 mg/d Q3: 2,280 mg/d Q4: 2,600 mg/d Q5: 3,240 mg/d

Follow-up Period	Health Outcome	Confounders Adjusted for	Results
6.3 y	Stomach cancer	Energy, age, sex, education level, self-reported stomach disorders, family history of stomach cancer, smoking status	No relationship between energy-adjusted salt intake quintiles and stomach cancer *Q1 vs. Q5* Positive, nonsignificant associations were found for bacon (RR=1.33; CI 1.03, 1.71) and other sliced cold meats (RR=1.29; CI: 0.96, 1.72, p for trend=0.07)

TABLE F-10 Evidence Tables: Gastrointestinal Cancer
Case-Control Studies

Citation	Population studied	Study Design	Sample Size (case/control)
Lazarević et al., 2011	Cases: Serbian, 45-85 y, diagnosed with gastric adenocarcinoma Controls: Serbians matched by age, sex, and residence	Hospital-based case-control	102 cases 204 controls
Lee et al., 2003	Cases: Korean men and women diagnosed with gastric cancer and without *H. pylori* infection Controls: Korean men and women	Hospital-based case-control	69 cases 199 controls
Peleteiro et al., 2011	Cases: Portuguese, diagnosed with gastric cancer Controls: Portuguese, 18-92 y	Hospital-based case-control	422 cases 649 controls

Sodium Exposure (method and level)	Health Outcome	Confounders Adjusted for	Results
98-item FFQ National food composition tables and USDA food composition tables Na intake tertiles (no ranges or median provided)	Gastric cancer		Association of Na intake with gastric cancer in men: *T2 vs. T1* *OR=4.66, CI: 0.28, 19.96, p=0.000 *T3 vs. T1* *OR=6.22, CI: 1.99, 7.86, p=0.000
Person-to-person interview conducted using semiquantitative 161-item FFQ No Na intake levels provided; frequency of consumption of various foods	Gastric cancer	Age, sex, family history, duration of education, smoking, drinking, *H. pylori* infection	Increase in early gastric cancer risk positively and significantly associated with increased intake of salt-fermented fish (*HR=2.4, CI: 1.0, 5.7) and kimchi (*HR=1.9, CI: 1.3, 2.8)
Semiquantitative 82 item FFQ T1: <3,067.5 mg/d T2: 3,067.5-3,960.1 mg/d T3: >,3960.1 mg/d	Gastric cancer	Age, sex, education, smoking, *H. pylori* infection, total energy intake	Risk of gastric cancer associated with highest salt exposure (T3 vs. T1): *OR=2.01, CI: 1.16, 3.46

continued

TABLE F-10 Continued

Citation	Population studied	Study Design	Sample Size (case/control)
Pelucchi et al., 2009	Cases: Italian men and women, 22-80 y, with confirmed stomach cancer Controls: Italian men and women, 22-80 y, frequency matched by age and sex	Hospital-based case-control	230 cases 547 controls
Strumylaite et al., 2006	Cases: Lithuanian with newly diagnosed gastric cancer, 22-86 y Controls: Lithuanian individually matched by gender and age ±5 y	Hospital-based case-control	379 cases 1,137 controls

Sodium Exposure (method and level)	Health Outcome	Confounders Adjusted for	Results
78-item FFQ grouped into 6 sections (milk/hot beverages, bread/cereal dishes, meat/main dishes, vegetables, fruit, sweets/ desserts/soft drinks) Na intake computed using an Italian food composition database (with other sources when needed) Na intake quartiles (not provided)	Gastric cancer	Education, period of interview, BMI, smoking, family history of stomach cancer in first-degree relatives, total energy intake	Gastric cancer was associated with Na intake compared to Q1: Q2: OR=2.22, CI: 1.27, 3.88 Q3: OR=2.56, CI: 1.41, 4.63 Q4: OR=2.46, CI: 1.22, 4.95 *p for trend=0.02
Self-administered structured questionnaire about dietary habits (56 diet items) based on the Aichi Cancer Center Questionnaire No Na intake levels provided	Gastric cancer	Smoking, alcohol consumption, family history of cancer, education level, residence, other dietary habits (e.g., speed of eating), other dietary habits, smoking, alcohol consumption, family history of cancer, education level, residence	Increased risk of gastric cancer associated with: Use of additional salt: *OR=2.98, CI: 2.15, 4.15, p for trend<0.001 Liking salty foods: *OR=3.88 CI: 1.98, 7.60, p for trend<0.001 Putting additional salt on prepared meal: *OR=2.98 CI: 2.15, 4.15, p for trend<0.001

continued

TABLE F-10 Continued

Citation	Population studied	Study Design	Sample Size (case/control)
Zhang and Zhang, 2011	Cases: Japanese men and women diagnosed with gastric cancer, 40-75 y Controls: Japanese men and women, 35-77 y	Population-based case-control	235 cases 410 controls

Sodium Exposure (method and level)	Health Outcome	Confounders Adjusted for	Results
98-item FFQ Daily Na intake calculated by national food composition tables and the USDA food composition tables Na intake tertiles T1: <3,000 mg/d T2: 3,000-5,000 mg/d T3: >5,000 mg/d	Gastric cancer	Age, sex, education level, smoking, alcohol intake, *H. pylori* infection	Na intake was associated with an increased risk of gastric cancer: T1: OR=1.00 T2: OR=1.95, CI: 1.23, 3.03, p=0.012 T3: OR=3.78, CI: 1.74, 5.44, p=0.12